The Biology of Grasslands

THE BIOLOGY OF HABITATS SERIES

This attractive series of concise, affordable texts provides an integrated overview of the design, physiology, and ecology of the biota in a given habitat, set in the context of the physical environment. Each book describes practical aspects of working within the habitat, detailing the sorts of studies which are possible. Management and conservation issues are also included. The series is intended for naturalists, students studying biological or environmental science, those beginning independent research, and professional biologists embarking on research in a new habitat.

The Biology of Streams and Rivers
Paul S. Giller and Björn Malmqvist

The Biology of Soft Shores and Estuaries
Colin Little

The Biology of the Deep Ocean
Peter Herring

The Biology of Soil
Richard D. Bardgett

The Biology of Polar Regions, 2nd Edition
David N. Thomas et al.

The Biology of Caves and Other Subterranean Habitats
David C. Culver and Tanja Pipan

The Biology of Alpine Habitats
Laszlo Nagy and Georg Grabherr

The Biology of Rocky Shores, 2nd Edition
Colin Little, Gray A. Williams, and Cynthia D. Trowbridge

The Biology of Disturbed Habitats
Lawrence R. Walker

The Biology of Freshwater Wetlands, 2nd Edition
Arnold G. van der Valk

The Biology of Peatlands, 2nd Edition
Håkan Rydin and John K. Jeglum

The Biology of African Savannahs, 2nd Edition
Bryan Shorrocks and William Bates

The Biology of Mangroves and Seagrasses, 3rd Edition
Peter J. Hogarth

The Biology of Deserts, 2nd Edition
David Ward

The Biology of Lakes and Ponds, 3rd Edition
Christer Brönmark and Lars-Anders Hansson

The Biology of Coral Reefs, 2nd Edition
Charles R. C. Sheppard, Simon K. Davy, Graham M. Pilling, and Nicholas A. J. Graham

The Biology of Grasslands
Brian J. Wilsey

The Biology of Grasslands

Brian J. Wilsey

Iowa State University, Ames, USA

OXFORD
UNIVERSITY PRESS

OXFORD
UNIVERSITY PRESS

Great Clarendon Street, Oxford, OX2 6DP,
United Kingdom

Oxford University Press is a department of the University of Oxford.
It furthers the University's objective of excellence in research, scholarship,
and education by publishing worldwide. Oxford is a registered trade mark of
Oxford University Press in the UK and in certain other countries

First Edition published in 2018
Impression: 1

Published in the United States of America by Oxford University Press
198 Madison Avenue, New York, NY 10016, United States of America

British Library Cataloguing in Publication Data
Data available

Library of Congress Control Number: 2018934542

ISBN 978-0-19-874451-1 (hpk.)
ISBN 978-0-19-874452-8 (pbk.)

DOI 10.1093/oso/9780198744511.001.0001

Printed in Great Britain by
Bell & Bain Ltd., Glasgow

Preface and Acknowledgments

Grasslands are vast and important to humans. Grasslands hold vast amounts of biodiversity, and it has been estimated that 3/5 of calories consumed by humans comes from grasses. Most grain consumed is produced in areas that were formerly grasslands or wetlands. Grasslands are also important because they are used to raise forage for livestock, biofuels, they sequester vast amounts of carbon, they are important as urban green-space, and for the maintenance of biodiversity. Intact grasslands contain an incredibly fascinating set of plants, animals, and microbes that have interested several generations of biologists. Grasslands have also been used to study important theoretical questions in ecology. Hopefully, this book will be helpful in providing an overview of the importance and interesting organisms and interactions within grasslands. I would like to thank the editor of this series, Ian Sherman, for suggesting that I write this book. My wife Anna Loan-Wilsey commented on some early chapters and Wayne Polley read and commented on all the chapters, and I thank them for their efforts. Emily Powers helped to compile Table 7.1. The target audience for this book includes advanced undergraduate and graduate students, postdoctoral fellows, senior researchers who are new to grassland systems, and members of the educated public who are interested in the biology of habitats.

My research focuses on how biodiversity and ecosystem functioning are related in grassland systems, how biodiversity-ecosystem functioning relationships are altered in novel ecosystems and with global change, and how restoration projects can be improved to increase biodiversity in heavily human impacted areas. I have conducted grassland studies in Tanzania, in central North America from Manitoba to central Texas, and in Yellowstone National Park. I have visited grasslands on all continents except Australia. I completed a Ph.D in grassland ecology under Sam McNaughton and Jim Coleman at Syracuse University, and have been a professor at Iowa State University since 2001.

Contents

1 Grasslands of the World

> I can sit on the porch before my door and see miles of the most beautiful prairie interwoven with groves of timber, surpassing, in my idea, the beauties of the sea. Think of seeing a tract of land on a slight incline covered with flowers and rich meadow grass for 12 to 20 miles...
>
> John Brooke, Early Texas settler, 1849, quoted in Diggs et al. (1999).

Grasslands are extensive across the globe and are found on every continent except Antarctica. What are grasslands? Grasslands are herbaceous dominated areas with very low abundance of trees and shrubs. They are usually dominated by grasses or other grass-like plant species, although they can also be surprisingly diverse in their numbers of non-grass wildflower species. They are found in areas with intermediate precipitation amounts (250–1000 mm) characterized by occasional droughts and are usually the most extensive in the interior of continents. The range of precipitation for grasslands is high, and they can easily become forested at the wet end of the spectrum, and deserts at the dry end of the spectrum, depending partially on human activities in the area. Grasslands can be temperate or tropical, and sometimes arctic tundra is considered a grassland type. Temperate grasslands have a continental climate, with typically hot summers and variable temperatures. Tropical grasslands have a wet and dry season.

Grasslands are hugely important to humans, as I will outline in this book. Grasslands are used to produce food (e.g., beef, mutton), energy (wind, biofuels), biodiversity conservation, carbon storage, and tourism. They are also used as a living genetic library, producing germplasms for future crop plants and ornamentals. Many common ornamental plants originated in grasslands. The Fertile Crescent region of the Middle East contained mostly steppe grassland and desert grassland when its inhabitants domesticated wheat, barley, peas (e.g., lentils), goats, sheep, and cattle. Many endemic birds and mammals are found in grasslands. Some of the most popular national parks in the world such as the Serengeti National Park in Tanzania and Kenya have grasslands, and grassland tourism can be a major source of

The Biology of Grasslands. Brian J. Wilsey. Published 2018 by Oxford University Press. © Brian J. Wilsey 2018.
DOI 10.1093/oso/9780198744511.001.0001

income to some countries. Grassland areas are also home to major cities such as Denver, Dallas-Fort Worth, and Nairobi.

1.1 Are grasslands human or climatically determined?

Because humans burned many grassland areas before European settlers arrived, some scientists in the past have viewed grasslands as purely anthropogenic artifacts. Bond (2008) noted that generations of ecologists have debated whether grasslands are anthropogenic in origin or not. Within some forested biomes, grasslands were created by humans clearing and burning forests in fairly recent and well documented time periods. These **human-derived grasslands** are 'early successional' systems that would not exist without human influences. However, several lines of evidence do not support the view that all grasslands are purely human artifacts. First, grasslands have been around for millions of years based on paleontological evidence (Edwards et al. 2010), i.e., millions of years before humans. Second, there are many endemic species in grasslands, especially in tropical and subtropical grasslands, which suggests that they are ancient. Third, grassland species have evolved many adaptations to fires and grazing, such as well-protected buds and storage reserves, resprouting ability, and flammable litter (Bond 2008). Very high growth in wet seasons and slow rates of decomposition result in large fuel accumulations (Bond 2008), which favor fire. This suggests that fires are ancient as well, which would have kept grasslands in an open state. Milchunas et al. (1988) categorize these types of grasslands as climatically determined, suggesting that they will persist without human intervention if they remain intact.

1.2 Origin of grasslands

Grasslands began to form 11–24 MYA when grasses invaded savannas and woodlands. Grasses initially all had C_3 photosynthesis and could have been found in shaded conditions. However, grasses with C_4 photosynthesis require full sun, and they increased in abundance during this time to achieve between 20–40 percent of the local vegetation. By around 6–8 MYA before the present, C_4 grasses were widespread on most continents that now have extensive grasslands. Edwards et al. (2010) reviewed the data collected on pollen, phytoliths, grazing mammal teeth, and stable carbon isotopic signatures and found that C_4 grass dominance, which indicates open sunny conditions, emerged at about 7, 7, 10, and 8 million years ago in Asia (China), Great Plains of North America, Africa (Kenya), and South America (Argentina). The area covered by grasslands has waxed and waned over time since then due to ice ages and interglacial periods during the Pleistocene. Ehleringer et al. (1997) argued that based on physiological considerations, specifically atmospheric CO_2 concentrations, C_4 grasses would have been more widely distributed during ice ages when CO_2 concentrations were at their lowest, than during interglacial periods.

1.3 Grassland soil types

Soil formation is a function of time, parent material, topography, climate, and organisms (Jenny 1980). Grasslands are found on a variety of soil types, from high clay soils to loams to sandy soils. Mollisols, Vertisols, and sandy Arenesols are grasslands soils. They are typically high in organic matter due to large root inputs and low shoot inputs and have low decomposition due to low water availability. Because of the high soil organic matter from the long history of litter inputs from deeply-rooted species, the soils make excellent farm land. Thus, many grassland areas have been converted to cereal agriculture; corn in tallgrass areas and wheat in shortgrass areas. These cereals are very important in supporting the large human populations that exist on Earth.

1.4 Grassland types based on precipitation and temperature

Ecosystems can be aligned by their relative levels of precipitation and evapotranspiration. Areas that have precipitation that exceeds evapotranspiration are humid and subhumid areas. **Humid grasslands** are found in areas that receive enough precipitation to support forests. **Subhumid** areas are more typically grasslands. Areas where precipitation does not exceed evapotranspiration are semi-arid and arid. **Semi-arid grasslands** are dry grasslands also called steppe. **Arid grasslands** are the driest and are sometimes considered to be desert systems.

1.5 Grassland types based on an anthropogenic gradient

Another way to categorize grasslands is to think about the boundary between grasslands and purely agricultural systems such as crop fields. Crop fields are not considered grasslands, but pastures are usually considered grasslands. Most grasslands in the world today are managed to various degrees, and some would not persist as grasslands without human intervention. Although some grasslands are ancient, they are certainly not all ancient in our human-dominated world of the present day. Many have been greatly altered by humans, and terms such as **'improved grasslands'** are used to describe human alterations. Improved grasslands are usually over-seeded with cultivated varieties of grasses and legumes and are fertilized and often sprayed with broad-leafed herbicides. They might not persist as simplified grasslands without human intervention. They differ fundamentally from wild unimproved grasslands.

Grasslands span an anthropogenic gradient from wild grasslands that are largely intact, to extensively managed grasslands, to intensively managed and simplified by humans (Figure 1.1). This gradient is due to the **intensification of agriculture**. On the anthropogenic end of the spectrum are single-species pastures with heavy inputs (e.g., fertilizers, herbicides). Biomass is harvested by livestock or by haying. These pastures will not persist on their own once human inputs are removed. Single-species pastures (monocultures) that are fertilized and sprayed with herbicide for weeds are usually considered to be perennial crop systems and not grasslands, but this view is not universal. If they are abandoned, these areas do not persist as monocultures, although they can become a type of modified grassland once they are invaded by other species. On the other end of the spectrum are large, intact grasslands that are found in areas that are not amenable to crop production. These areas may still have significant levels of grazing by native herbivores. Anthropogenic intermediates, sometimes called **degraded grasslands** are intact grasslands have been overseeded with non-native plant species, usually cultivars, have had excessive grazing by livestock, or have been altered in some way. These pasture grasslands are very important for beef and lamb

Remnant Grasslands:	Intermediate Grasslands:	Cropland pastures:
Prairie		
Pampas	Haying	Herbicides
Veldts	Mowing/threshing	Fertilizer
Rangelands	Introduced species	Cultivars of
Steppe	Predator removal	introduced species
Puszta	Managed grazing	Monocultures
Natural fires	Fire exclusion	

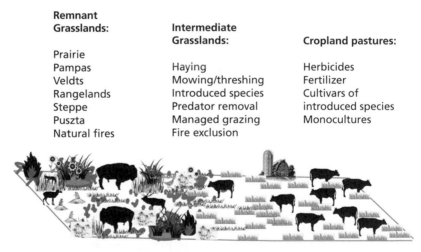

Figure 1.1   Anthropogenic gradient across grassland types, with human impact increasing moving from left to right. Wild, never plowed grasslands are on the left, which have native species at multiple trophic levels, patchy fire and grazing, and diverse mixture of animals present. They are usually called remnant or relict grasslands. On the right are the most heavily human impacted grasslands, which typically have a single domestic species, a simplified and low diversity plant community, and heavy human inputs of fertilizer and pesticides. They are typically called improved pastures. These grassland types can function quite differently from one another.

production but can be hard to restore to native grassland when they are abandoned. This mixture of grassland types depending on the level of human impacts makes grasslands challenging, but interesting to study (Figure 1.1). An important question facing humans in many countries is how to produce protein for human consumption, which sometimes requires intensification, while maintaining high biodiversity and ecosystem services in the environment by protecting wild areas.

An associated human impact in modified grasslands is animal husbandry. In some intact livestock-grazed grasslands, animals are vaccinated and fed hay during time periods of low food availability. This results in pastures having an order of magnitude greater consumption rates than wild grasslands (Oesterheld et al. 1992). These higher consumption rates make comparisons of wild herbivore vs. livestock-based grazing systems difficult.

1.6 How extensive are grasslands globally?

The boundaries between grasslands and other ecosystems can be poorly delineated, and as a result, estimates of the proportion of the earth's surface covered by grasslands can vary considerably among researchers (Table 1.1). The area covered by grasslands has shrank and expanded over time due to human activities, but is now considered to be about 40 percent of the Earth's surface when tundra and some shrublands are included (Table 1.1). Excluding tundra and shrublands leads to an estimate of about 20–25 percent of the Earth's surface in grasslands. In the high rainfall part of the spectrum, grasslands are found only when there is frequent fire or disturbances by humans or

Table 1.1 Estimates of percentage of the earth's surface that is grassland.

Grassland Type	Coverage	Total	Total without tundra	Total without shrublands
Whittaker and Likens (1973)		31.4%	25.2	24.8
Temperate Grassland	7.0%			
Savanna	11.6%			
Dry Savanna/Woodland	6.6%			
Tundra	6.2%			
Lieth (1975)		24.3%	18.3%	
Grassland	18.3%			
Tundra	6.0%			
Atjay et al. (1979)		34.4%	27.1%	31.7%
Temperate Grassland	9.7%			
Savanna	9.3%			

(continued)

Table 1.1 (Continued)

Grassland Type	Coverage	Total	Total without tundra	Total without shrublands
Dry Savanna/Woodland	2.7%			
Tundra	7.3%			
Matthews (1983)		26.2%	20.7%	
Grassland	20.7%			
Tundra	5.5%			
Olson et al. (1983)		55.5%	41.9%	42.3%
Non-woody Grassland or shrubland	21.4%			
Savanna	7.3%			
Dry Savanna/Woodland	13.2%			
Tundra	13.6%			
Pilot Analysis of Global Ecosystems (2000)		40.5%	34.8%	27.8%
Savanna	13.8%			
Non-woody Grassland or savanna	13.8%			
Shrublands	12.7%			
Tundra	5.7%			

Total ice-free land surface area is 129,476,000 km²; so, to convert to area, multiply X%/100 x area.

Figure 1.2   Savanna in Tanzania, Africa. Photo: Brian Wilsey.

herbivorous mammals. They grade into deserts at the dry end of the spectrum. **Savannahs** are open grasslands with scattered trees (Figure 1.2). They are extensive in Africa and the Cerrado of South America, and in other continents at the forest-grassland boundary. Relatively open savannahs will be considered here as a grassland type, but they can grade into a closed canopy savannah, which is a woodland type. As the proportion of trees increases, sites are typically called grassland, savannah, woodland, and then forest.

1.7 Continents with major grassland areas

1.7.1 Europe

Much of Europe is dominated by human derived grasslands (Hejcman et al. 2013). These grasslands are maintained by sheep and cattle, sometimes rabbits, and a lack of forest seed sources (Figure 1.3). In England, the removal of these herbivores and human inputs does not lead to tree establishment without tree seed additions, which suggests that the current vegetation state is stable (Tansley and Adamson 1925). The formation of human-derived grasslands has increased the amount of land in grassland world-wide, and this has countered some of the loss from woody plant encroachment and conversion of the original grasslands to croplands. However, not all European grasslands are human derived, especially in Eastern Europe. The Puszta is a large intact grassland in Hungary that is climatically determined and is not human derived.

Many human-derived grasslands have been grasslands since the middle ages. Calcareous grasslands are high pH areas found in the UK, Germany, and other parts of Europe (Figure 1.3). Detailed laws require landowners to hay their lands during late summer to prevent woody plant encroachment to maintain their grassland status in many countries. In Germany, many land-owners are even required to use hand tools in fen areas.

Figure 1.3   Southdowns chalk cliff grassland, southern England. The shallow soils are of Cretaceous origin, when the area was below sea level and Coccolithophores (sea organisms) where present. The area was uplifted in the Cenozoic and was eroded during the last ice age. Grazing by sheep and browsing by rabbits over many centuries has kept it in an open state. Photo: Anna Loan-Wilsey, used with permission.

Hejcman et al. (2013) reviewed that history of Central European grass-lands from data on non-arboreal/arboreal pollen ratios, charcoal in sediments, and the soil profile. Non-arboreal pollen and charcoal are both indicative of a grassland state. They classified grasslands into three types: 1) natural grasslands, predetermined by environmental conditions, and wild herbivores, 2) semi-natural grasslands associated with long-term human activity from the beginning of agriculture during the Mesolithic-Neolithic transition, and 3) improved (intensive) grasslands sown with highly productive forage grasses and legumes. Improved grasslands have increased in area in recent times. Historically, natural grasslands were present as small grassland patches that were fragments embedded in the forested landscape prior to 5,500 BC. The introduction of the scythe in the seventh to sixth centuries BC led to the enlargement of grasslands for hay meadows, and paleontological evidence for extensive grasslands in Europe is found by the Middle Ages. Since that time, many hay meadows have been maintained in an open state.

During recent times, the proportion of grasslands in Europe has been more stable (Hejcman et al. 2013). During war times, grasslands are converted into crop fields, and crop fields were converted back into grasslands after the collapse of agriculture in former communist countries in the 1990s.

Many European grasslands were historically grazed by the auroch, a species native to Europe. The auroch was the descendent of modern-day domestic cattle. They went extinct in 1627, and little is known about their original habitat. The last remaining habitats for auroch before they went extinct were marshy areas and forests, but this may have been because human activity was lowest in these areas. They were grazing mammals, and presumably were widespread in open grassland and wetland environments in Europe. A current effort to breed cattle for a greater proportion of auroch genes is ongoing, with the hope to reintroduce these auroch-like cattle into some locations of Europe.

1.7.2 Asia

The Eurasian steppe system extends from Eastern Europe (Romania) through Ukraine, Mongolia, Russia, to northeast China. Europe and Asia are separated by the Ural Mountains, and this provides a barrier for plants, microbes, and some small animals, but not for some large mammals. The southern portion of Europe and Asia are connected by the Mediterranean Sea, which makes the large mammal (especially predators) composition more similar between Europe and Asia than it is for plants and other organisms. It has also become more homogenized over time due to human movements between Europe and Asia.

The Asian steppe is one of the largest grasslands in the world (Figure 1.4). In China, steppe grasslands extend from the northeast to the southwest, ending in Tibetan montane regions, covering ~40 percent of the land area (Wang and Ba 2008). The dominant plant species are the grasses *Leymus chinensis* and species of *Stipa*. Alfalfa (*Medicago sativa*) is a forb native to Chinese grasslands. Alfalfa is now one of the most important forage crop plants throughout the world.

Originally, Asian steppes supported several native herbivore species, many of which are now rare or threatened with extinction: a true wild horse—the Przewalski's horse (*Equus ferus*), saiga antelope, Mongolian gazelle, the Asiatic wild ass, and the Bactrian camel. These native herbivores have been replaced with domestic species in most areas. The principal grazing mammals now are domestic sheep, goats, horses, and cattle. Wolves were major predators that were extirpated in most areas.

Many Asian steppe grasslands are highly degraded due to chronic high grazing and too frequent mowing for hay (Wang and Ba 2008). The grazing regime in Asian steppe has changed over time as people settled and left behind a nomadic way of life. Many Asian grasslands evolved with grazing and frequent moving by nomadic bands of people and livestock, which helped to distribute grazing pressure (Sneath 1998). With settlement, this pattern of grazing changed, and the stocking rate increased with the rise of human populations. Many areas of the Inner Mongolian region of China are now degraded from intense grazing (Wang and Ba 2008), and these grasslands contain extensive alkaline and saline bare ground areas with high soil pH (up to 9.0, Figure 1.4).

Figure 1.4   A meadow grassland in the Eurasian steppe region of China. Pictured left to right are the author Brian Wilsey and Deli Wang. The common grass is *Leymus chinensis*, a dominant species in many Chinese grasslands. This is a nutrient poor habitat, with soil of high salinity and alkalization (pH 8.5–10.0). Photo: Deli Wang used with permission.

The meadow steppe of China is well studied and can be classified into three types: bunchgrass meadow steppe, rhizome grass meadow steppe, and forb meadow steppe. The bunchgrass and forb meadow steppe are the most diverse, containing 17–27 species m^{-2} (Wang and Ba 2008). Primary productivity is 140–350 g/m^2.

1.7.3 Australia and New Zealand

Grasslands in Australia are found under highly variable conditions: from 11°S to 44°S in latitude, and receiving 100–4000 mm in precipitation (McIvor 2005). Grasslands at the high end of the precipitation spectrum are human derived grasslands. A total of 70 percent of the continent is used for foraging, most commonly by sheep and cattle. Common wild animals are kangaroos and wallabies, and the exotic species the European rabbit. In evolutionary times, the grazing intensity was assumed to be lighter in Australia than in East Africa or central North America (McIvor 2005). Moore (1970) classified Australian grasslands into 13 types, with the 13 types falling into three broader groups of tall grasslands (>900 mm), midgrass areas (450–900 mm) and shortgrass areas (<450 mm). Many exotic plant species were introduced to support domestic grazing animals after European settlement (Firn et al. 2012), and the African C$_4$ grass *Themeda triandra* is now the most widely distributed of all plant species in Australia (Gallagher 2016). New Zealand has many grassland types, and interestingly, did not have mammalian grazing species prior to human settlement.

1.7.4 North America

Grasslands are extensive in the Great Plains region of central USA and Canada. They are also common in isolated pockets within forest biomes in the eastern USA and Canada, in the northwest USA (Palouse prairie), and in California, and Oregon. Central USA grasslands are classified according to the height of the dominant grasses, with **tallgrass prairies** extending from Manitoba south in a band through central Texas. An archipelago of tallgrass prairie extends from Minnesota and Iowa east through Illinois into parts of Indiana and Ohio, called the **Prairie Peninsula**. The tallgrass prairies have been extensively lost to agriculture, primarily to corn and pasture, with up to 99.9 percent converted in some states (Samson and Knopf 1994), and >90 percent overall. Tallgrass prairies that are remaining are very diverse, with common plant species being big bluestem *Andropogon gerardii*, indian grass *Sorghastrum nutans*, switchgrass *Panicum virgatum*, little bluestem *Schizachyrium scoparium*, dropseeds (*Sporobolus* spp.), and many wildflowers such as asters, blazing stars (*Liatris* spp.), coneflowers (*Echinacea* spp.), sunflowers (*Helianthus* spp.), prairie clovers, and other legumes (Tilman et al. 2001; Knapp et al. 2002; Wilsey et al. 2005). Sunflowers, which are native to North America, are now major oil and seed producing crops grown world-wide, and coneflowers and blazing stars are important landscaping plants. Tallgrass prairies are found at the uppermost limit for grassland precipitation and can convert to forest with the cessation of burning and grazing.

Shortgrass prairie, or **shortgrass steppe** extends from Calgary in the north through Montana, Wyoming, and Colorado, ending in the panhandle of Texas. It is dominated by blue gramma (*Bouteloua gracilis*) and buffalo grass (*Buchloe dactyloides*), and occasionally sagebrush (*Artemisia* spp.). **Mixed grass prairie**, which contains a mixture of short and tall grass species, extends from the Dakotas south to west Texas in between the tallgrass and shortgrass regions. The beautiful sand hills of Nebraska are mixed grass prairie. The mixed grass and shortgrass prairie systems are much more intact compared to tallgrass prairie. Arid grasslands abut desert systems in the southwestern USA and Mexico and can be quite extensive in New Mexico and Arizona.

Grasslands are found in other parts of North America in less extensive tracts. California grasslands are dominated by annual plant species and typically have a Mediterranean climate with wet winters and dry summers. The Palouse Prairie was originally found in Washington and Oregon, and now exists as small remnants surrounded by wheat fields. Small isolated prairies are also found in the southeastern USA (Weiher et al. 2004). Human derived grasslands are extensive in the eastern USA and Canada.

1.7.5 South America

Although South America is famous for having one of the largest intact tropical forests in the world in the Amazon basin, it also contains extensive grasslands. The major grassland regions are the *Pampas* of Argentina, the *Campos* of Uruguay and Brazil, and the *Cerrado*, a savanna region of Brazil. The *Pampas* and *Campos* grasslands are variable, with a mean annual T of 10–20°C and receive 400–1600 mm precipitation (semi-arid to humid). Temperate sub-humid grasslands are found from 28–38°S in the eastern part of South America (Soriano 1991). All of these systems are important in raising livestock for beef and mutton production.

The *Pampas* are relatively flat to rolling, found on Mollisol soils, and are important as areas of cattle production (Figure 1.5) (Tognetti et al. 2010). The region is fairly diverse, with 1000 vascular plant species present (Soriano 1991). The *Pampas* were lightly grazed for millennia until the arrival of Europeans in the sixteenth century, and the abundance of grazing mammals varied more widely over geological time than grasslands in other regions (Edwards et al. 2010). The principal large mammalian herbivore was the Pampas deer (*Ozotoceros bezoarticus*). Rhea (or ñandu, *Rhea americana*) are large flightless birds that were originally fairly common omnivores that did some grazing. Rhea are less common now because of wire fencing and uncontrolled hunting. The Pampas also had many Plains Viscacha (*Lagostomus maximus*),

Figure 1.5   Exclosure in the flooding Pampas region of Argentina. Area to the left excluded cattle, and the area to the right allowed cattle grazing. The Pampas of South America are major cattle producing regions. Photo: Brian Wilsey.

which are rodents in Chinchilla family that build large underground colonies, similar to prairie dogs of North America. Plains Viscacha are commonly exterminated by ranchers due to their dietary overlap with cattle (Pereira et al. 2003).

1.7.6 Africa

Most African grasslands have scattered trees in them and are called savanna. Most trees are in the genus *Acacia* and have a characteristic flat-topped architecture (Figure 1.6). The density of trees can vary greatly and can be near zero in areas such as the Serengeti Plains and very high in heavily wooded areas called woodlands or forests. The Greater Serengeti Ecosystem in East Africa is one of the most well-studied grassland-savanna areas of the world. Across a large gradient of sites, Serengeti grasslands can be defined as short-grass grassland, mid-grass grassland, and tallgrass grassland/savanna. African grasslands contain some of the most spectacular wild animal populations in the world, including animal groups like elephants that have gone extinct in other parts of the world during the Pleistocene.

Most African grasslands and savannas have had a long history of relatively intense grazing by a variety of herbivores. The effects of grazing mammals will be covered in future chapters. The intensity of grazing has led to the human collection of African plant species for reintroduction elsewhere as forage plants, such as *Cynodon dactylon* (Bermuda grass), *Hyparrhenia rufa* (Figure 7.5), *Themeda triandra* (Red oat grass), and *Panicum coloratum* (Klein grass). Many of these species are now dominant in grazed grasslands of North and South America and Australia.

Arid grasslands in northern Africa are converting to desert ('desertification') in some areas. Desertification occurs when livestock grazing is conducted

Figure 1.6   Flat-topped Acacia tree in savanna of Tanzania. Photo Brian Wilsey.

during inappropriate times causing plant mortality. Recruitment from seed is low and topsoil loss due to erosion is high after the dominant grasses die (Schlesinger et al. 1990). Grazing by cattle in semi-arid grasslands can lead to increases in the spatial and temporal heterogeneity of water and soil nutrients (Schlesinger et al. 1990). After grazing, the concentration of soil N is much higher under shrubs than in the barren area between shrubs, and this positive feedback results in the conversion of semi-arid grassland to desert (Schlesinger et al. 1990). The authors of this paper speculated that this process is operating in the Sahel as well as the Southwest USA. This process may be exacerbated with increases in atmospheric temperatures expected in the future.

1.8 Human evolution and biophilia

African savannahs are also important because it is where *Homo sapiens* (humans) speciated and eventually spread to other continents. A 1.9 million-year-old *Homo habilis* was found by the Leakey's at Olduvai Gorge in

Figure 1.7   Olduvai Gorge, Tanzania (Note: misspelling of Oldupai, which is the wild sisal plant. The gorge was named after an early European explorer asked a Maasai herdsman the name of the gorge, and the Maasai thought he was asking the name of the plant.) This area, which borders the Serengeti Plains region, was the location of many findings of early hominid fossils by Louis and Mary Leakey, including *Homo habilis* and several Australopithecines. Photo: Brian Wilsey.

Tanzania (Figure 1.7). They have also been found in Ethiopia. From Africa, humans have spread and established populations on all continents, although the population at Antarctica is perhaps not permanent.

Some have hypothesized that our shared evolutionary history has led to interesting subconscious preferences for savanna-like structures around homesteads and in city parks (Wilson 1992). Humans create parks in forested regions by clearing trees until they are scattered individuals with short grass between them. In non-forested areas, humans plant trees so that they are more "park-like", that is, with scattered individuals with mowed grass underneath. Mowed grassy areas make us feel content because they look similar to grazing lawns, which subconsciously makes us feel like there is plenty of food in the area (McNaughton 1984). Trees are typically trimmed in parks so that they do not have lower branches, which is typical of savanna areas with fire and browsing present (Wilson 1992). Our shared evolutionary history has led to our instinctive preference for open areas with scattered trees on hilltops above permanent water bodies. This might explain several social factors involved in grassland ecology, such as human tolerance for some woody plant encroachment into grasslands and clearing of forests.

2 Biodiversity of Grasslands

Because grasslands are herbaceous (non-woody) plant dominated systems, they lack the tree-understory structural differences of forested systems. Nevertheless, grasslands can be surprisingly diverse and contain many charismatic flora and fauna. In grasslands, large mammals fill the niches that are filled by insects and birds in forests.

2.1 Flora

The most abundant plant species in grasslands, as their name suggests, are grasses. There is a surprisingly large amount of variation among grass species. Grass species can be rhizomatous, stoloniferous, or caespitose (bunchgrasses), can have C_3 or C_4 photosynthesis, can be tall or short, and can be members of multiple clades within the grass family (Poaceae) with a different evolutionary history. They can be **annuals** that grow and die in one growing season, or **perennials** that grow over two or more growing seasons. Grasses have important morphologies that explain their response to the environment. The crown of a grass is a storage organ (Figure 3.1) just below the soil surface; the roots and new shoots emerge from the crown. The crown is not affected by grazing on shoots and is usually not affected by fires. Fires typically do not penetrate very deeply into the soil. The roots that extend downward from the crown are fibrous as they are in all monocots (grasses, sedges, and rushes in grasslands). Perennial grasses can be very deeply rooted, with roots extending to 1 meter or deeper (Figure 2.1).

Perennial grasses also vary in the types of stems that they have and are often clonal in nature. Grasses can have underground stems called rhizomes that extend from the plant into new territory, or they can spread by aboveground horizontal stems called stolons. Many forbs are also clonal, and have rhizomes connecting plants belowground (Figures 2.2, 2.3). A new shoot (ramet) with its own root system can develop at the end of the rhizome or stolon, and these ramets have varying degrees of independence from other parts of

The Biology of Grasslands. Brian J. Wilsey. Published 2018 by Oxford University Press. © Brian J. Wilsey 2018.
DOI 10.1093/oso/9780198744511.001.0001

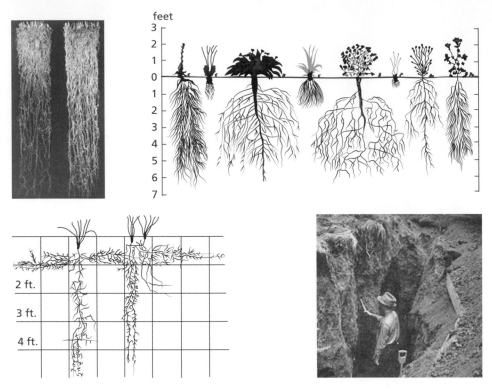

Figure 2.1   John Weaver and students excavated tunnels and estimated rooting depth of numerous grassland plant species in the early 1900s. From top left to bottom right: roots of big bluestem (*Andropogon gerardii*, left) and switchgrass (*Panicum virgatum*, right) to 150 cm (5 feet) deep (Weaver 1968), representative root systems from short grass prairie based on 325 root systems of these species, blowout grass *Redfieldia flexuosa* in Nebraska, and John Weaver in a trench (photos public domain). Weaver was a Victorian and went into the trenches dressed in a hat and tie. Data from Weaver's studies have been digitized and analyzed with GIS by Sun et al. (1997).

the plant. Grasses typically have rhizomes or stolons but not both, but two common species are exceptions to the rule (*Cynodon dactylon* and *Buchloe dactyloides*). Clonality has been found to be associated with a grasses ability to reduce local species richness, especially when it is tall and non-native (Dickson et al. 2014). Bunchgrass (caespitose) species grow in dense clusters, usually in a round bunch that spreads outward as the plant grows. In a common North American species, little bluestem (*Schizachyrium scoparium*), the innermost part of the plant can die leaving a donut-shaped ring.

There are major differences among grasses in mode of photosynthesis and height, and these differences are often used to classify grassland types. Grasses with C_3 photosynthesis ("cool-season" or C_3 grasses) and with C_4 photosynthesis ("warm-season" or C_4 grasses) differ in when they actively

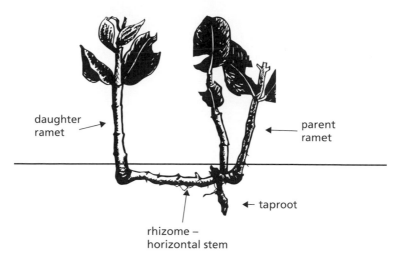

Figure 2.2   An example of a clonal plant (the forb common milkweed *Asclepias syriaca*). Many grass-land species are clonal and consist of a genet (genetically identical individual), and multiple ramets per genet. Horizontal stems can be above (stolons) or belowground (rhizomes, shown here). Connections between ramets can be severed by hooves or soil cracks, lead-ing to vegetative propagules becoming separate from the parent plant (vegetative repro-duction). Common milkweed is a host plant for the monarch butterfly.

grow (spring and fall, and summer in the Northern Hemisphere, respectively), and in resource use. These differences will be described in detail in Chapter 7. Grass species range in height from <20 cm to 2 or more meters, and height is at least partially genetically controlled (McNaughton 1984). **Tallgrass** systems (e.g., tallgrass prairie) are dominated by tall species, **shortgrass** systems being dominated by short species (e.g., shortgrass prairie or steppe), and **mixed (or midgrass)** systems are dominated by a mixture of both.

Grass clades, especially at the level of the 'Tribe', are important in under-standing the history of an area. The Panicoideae tribe for example, tends to be highly successful compared to other tribes for reasons that are poorly known (Edwards et al. 2010), and includes the biofuel species switchgrass (*Panicum virgatum*). The Andropogonidae tribe contains many of the "bluestem" species that are so common in many areas.

There are other monocot plant families that are similar to grasses, the sedges Cyperaceae, and the rushes Juncaceae. Sedges, rushes, and grasses are col-lectively called **"graminoids"**. Sedges are widespread and species-rich in grasslands. They differ from grasses in that they have triangular shaped stems. Rushes have round stems and are more abundant in wet areas within grasslands. Humans have used common names to label species in grass-lands, and they can sometimes be misleading. For example, "sawgrass" in the Everglades is actually a sedge, and bulrushes (*Scirpus*) are actually

Figure 2.3   An example of a clonal grass, in this case *Digitaria macroblephara*. Ramets, which can have their own root systems (three pictured here) are connected by stolons, or above-ground horizontal stems. Rhizomatous species have connected ramets connected by underground stems called rhizomes. Each ramet has several tillers. Notice that there is a reflection of the plant on the hood of the truck. Photo: Brian Wilsey.

sedges. Graminoids are members of the Monocot clade, and have narrow leaves with parallel leave veins, one seed leaf (cotyledon), and fibrous roots.

There are also many Eudicots in grasslands. Eudicots have broad leaves with palmate leaf venation, two seed leaves, and root systems that are not fibrous (usually tap-rooted). Eudicots are commonly called **forbs** by grassland ecologists, and they contribute to the great beauty of grasslands as wildflowers. They have also produced many ornamental garden plants. Forb species have a variety of flower colors, ranging from red to purple to blue to white and yellow. Most are animal pollinated and produce nectar. Thus, they attract a variety of pollinating bees and nectar-feeding butterflies. A few species are wind-pollinated (e.g., *Artemisia* or 'sage', *Plantago* spp. and false boneset *Brickellia eupataroides*) and contain green, non-conspicuous flowers. Many eudicot species are members of the Asteraceae family, but a multitude of other families can be present in diverse grassland areas. Members of the pea family (Fabaceae), are usually considered separately from other Eudicots due their association with nitrogen-fixing bacteria. Nitrogen-fixation occurs by bacteria in root nodules and involves converting atmospheric forms of N (N_2) into NH_3, which becomes NH_4 and NO_3, N forms that can be taken up by other plant species. Members of the Fabaceae family are usually called **Leguminous forbs** by grassland ecologists.

Because there are thousands of plant species in grasslands worldwide, and because there is a need to compare grassland sites in terms of their similarities and differences, plant species are often combined into 'functional groups' (similar to 'guilds' studied by animal biologists). Most commonly, we use the term 'graminoids' to refer to grasses and grass-like plants (sedges and rushes), 'forbs' for Eudicots (and some monocots), and 'leguminous forbs' for members of the pea family. Graminoids are commonly split into groups based on their mode of photosynthesis (C_4 and C_3). These groups can in turn be split into annual and perennial groups depending on their life-history status. Tilman et al. (1997) found that the number of functional groups was important to aboveground plant productivity in Minnesota grasslands, and that the number of functional groups was more strongly related to productivity than the number of species alone. Kindscher et al. (1995) compared plant species in a Kansas USA grassland and found that members of each of these functional groups clustered together based on a number of traits, which provided support for the concept of functional groups.

However, the forb group is especially heterogenous and should be split into multiple groups based on when species' actively grow and flower, whether they are **perennial** (living multiple growing seasons), **biennial** (two growing seasons, usually emerging as a basal rosette in the fall, and then flowering in the following spring), or **annual** (living one growing season), and based on rooting systems and clonal growth forms (Figure 2.4). Kindscher et al. (1995) found support for an 'early' and 'late' flowering forb group. Forbs are also commonly combined into groups based on heights, seed mass, maximum

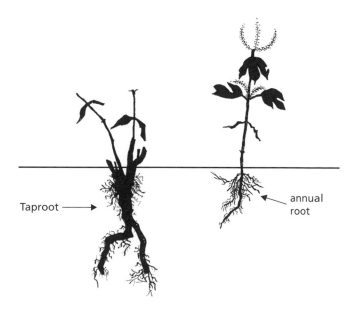

Taproot ⟶

⟵ annual root

Figure 2.4 Typical tap root systems in forbs. A taproot in a perennial species (on left, *Verbena stricta*, hoary vervain), and an annual root system in the annual species giant ragweed (*Ambrosia trifida*).

growth rates, and resource capture, specific leaf area (area/mass), and rooting depths (Grime 1977, Tilman 1982, Westoby 1998, Harpole and Tilman 2006, Dornbush and Wilsey 2010).

2.2 Plant syndromes and classification schemes

Three major conceptual models commonly are used to classify grassland plant species into a small number of groups, the competitors-stress tolerants-ruderals (CSR) grouping of Grime (1977), the leaf traits, plant height, seed mass (LHS) grouping of Westoby (1998), and the R* scheme of Tilman (1982). All three of these approaches were developed with, or extended to, grassland species.

Grime (1977) suggested that the **CSR grouping** could serve as a periodic table for ecology. Grime's framework (Figure 2.5) is based on a three-way trade-off between growth, storage, and sexual reproduction. In stressful environments, low resource availability would select for storage of resources and against growth (Chapin 1980, Chapin et al. 1990). Plants from stressful environments can have low growth rates even when fertilized, and their low growth rate in "good years" is puzzling unless you consider the trade-off between growth and storage. During times of plentiful water

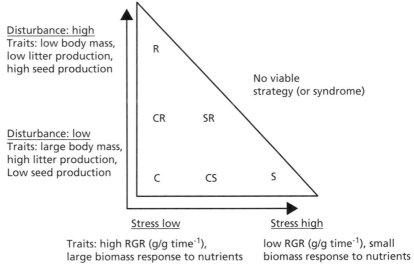

Figure 2.5 Grime (1977) suggested that grassland plant species can be classified into functional groups according to their relative growth rates, response to nutrient enrichment (Chapin 1980), and body mass and seed production rate. Stress is defined as an environmental factor that lowers growth rate, and disturbance is any factor that damages biomass. Many European grassland species have been classified according to this framework.

and nutrients, **S species** typically store the water and nutrients rather than allocating the resources towards growth (Chapin 1980, Chapin et al. 1990). They typically tolerate low-rainfall years by having smaller reductions in biomass production than other species. The presence of stress-tolerant species can reduce temporal variation in net primary productivity (NPP) across years (Polley et al. 2007, 2013). Competitive **C species** are similar to K selected species in that they have high body mass, produce lots of litter and many clonal ramets, and are competitively superior to other species in non-stressful, undisturbed environments. **R species** are similar to r-selected species in that they have short life spans, quick recovery from disturbance, and high seed production. The proportion of these groups in the community is predicted to vary across large environmental gradients, and competition (ecological interactions between species) is predicted to be more intense in benign environments than in stressful environments. Facilitation (+ + ecological interactions) is found more commonly in stressful environments like arid grasslands, deserts, and tundra, where plant growth is improved by the presences of neighboring plants that reduce stress by reducing wind or increasing soil oxygen concentrations (Bertness and Callaway 1994).

Grime's model is probably better as a predictor of grassland and desert plant species differences than tree and animal species. Grime's approach is important because it led to the huge interest in using traits to predict ecological impacts of species and their responses to environmental changes (Grime et al. 1997).

Westoby's (1998) LHS classification system predicts that leaf traits (L) underlying growth rates, especially specific leaf area (area of a leaf in cm^2 divided by mass), plant height (H), and seed mass (S), can be used to predict plant species impacts and responses (Figure 2.6). Specific leaf area is higher in plants with high photosynthetic and growth rates, and this measure is correlated with a wide variety of plant and ecosystem responses, including year to year biomass stability (Polley et al. 2013). Plant height (H) is an extremely important measure in grasslands, and has been used to classify grasslands into tall, mid, and shortgrass systems. It also frequently predicts the outcome of plant competition in undisturbed environments (Gaudet and Keddy 1988), and responses to grazing (McNaughton 1984). Herbivory to a given height will remove a greater proportion of biomass from a tall than shorter species. This can lead to shorter species being released from competition after a grazing event and can lead to having high abundance of short species in heavily grazed areas. Finally, seed mass (S) is important, especially during the regeneration phase of the plant life cycle. Large seed mass species are more likely to establish in areas with established vegetation (e.g., from overseeding) than small seed mass species. However, small seed mass species can disperse more widely, which underlies the competition-colonization tradeoff that is seen in some grasslands.

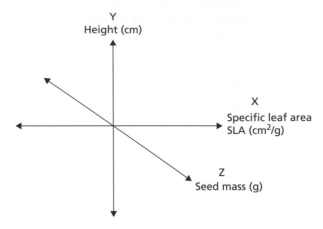

Figure 2.6   Westoby (1998) plant classification scheme. Species with high specific leaf area (SLA) have high photosynthetic and growth rates (X axis). Tall species are expected to have different ecological effects than short species (Y axis), and grasslands are classified according to plant height of dominant species. Species with large seed mass are predicted to be able to establish in dense vegetation and be more competitive than small-seed-mass species. Species with small seeds are predicted to produce more seeds per plant and to be better at dispersal than species with large seeds.

Tilman's R* approach was originally developed for algal communities, but was then extended to grasslands of Minnesota, USA. The assumption of the R* approach is that competition is important in regulating species abundances, and R is the most limiting resource in the environment. R* is the level to which a plant species can reduce a limiting resource, in most cases the soil nutrient N, via plant uptake. The species with the lowest R* will win at competition, because that species can reduce the resource to a low level that it, but not other species, can tolerate. The R* model predicts that competition intensity will not vary along stress gradients but will rather shift from aboveground competition to belowground competition as soil resources decline along gradients. Harpole and Tilman (2006) calculated R* for 27 grass and forb species in Minnesota and Kansas, and found that R* was a strong predictor of grassland species abundances. Species with low R* were more abundant than species with high R* in experimental plots, in old fields, and in intact prairies.

2.3 Fauna

The animal species in grasslands elicit feelings for the old west in the USA, or of the 'bush' in wild areas in Australia and Africa. Grasslands in many parts of the world once supported large populations of plant-eating mammals (herbivores). These species in turn supported carnivores and a well-developed invertebrate soil food web. These native herbivore assemblages have been replaced by domestic livestock to support food production in many grasslands of the world.

2.3.1 Large mammals

Herbivores all consume plants, but they can be classified into several functional groups that differ in their feeding ecology. Large herbivorous mammals can be classified into 'grazers', 'browsers' or 'mixed feeders' depending on what plants they feed on, and these groups can be delineated with C isotope fractions in bones and teeth based on the proportion of shrubs and trees and grasses in their diets (Ambrose and DeNiro 1986). The majority of the diet of **grazers** consists of grasses and other graminoids. Grazers are also called bulk and roughage feeders by Hofmann (1973). Because they feed on grasses and sedges, and because grasses and sedges are commonly dominant, grazers can have important effects on grasslands, including increasing nutrient cycling by converting grass tissues with low N concentrations to materials (urine and dung) with high N concentrations, causing a release of forbs from grass competition, which can lead to greater forb abundance and increased plant diversity, and altered C cycling (see Chapter 4 for more details). Graminoids have silica as a defense against herbivory, and this works against many insects but not all grazing mammals. Grazing mammals have molars that can crush and process the silica. Bison in North American (American buffalo), buffalo in Africa and Asia, domestic cattle, large antelope species such as wildebeest in Africa, and kangaroos in Australia are examples of important grazing species. Birds such as Rhea in South America and snow geese in wet grasslands are also grazers, although Rhea are omnivores that feed on small animals as well. The majority of the diet of **browsers** consists of forbs and woody plant shoots (browse). They are also called 'concentrate feeders' by Hofmann (1973). Browsers sometimes eat some grass, especially when it is young and green, but grass does not make up the majority of their diet. White-tailed deer in North and South America, and giraffes and small antelope species are important browsers. Browsers can have opposite effects on plant diversity than grazers due to their feeding on subordinate (forb) species rather than dominant (grass) species. Browsers usually selectively feed on nutritious young plant tissues and species, and can cause a decline in rare plant species when abundant (Waller et al. 1997, Côté et al. 2004). Browsers tend to have smaller body mass than grazers (Gordon and Illius 1994), with giraffes as major exceptions to the rule (Figure 2.7)!

Mixed feeders (sometimes call intermediate feeders) graze and browse, although this feeding activity can be separated in time. Many mixed feeders will graze when graminoid tissues are young and nutritious but will shift to browsing when grasses are dormant or stemmy. Elk (*Cervuus canadensis*), eland, impala, and African elephants are classic mixed feeders. Domestic

Figure 2.7   An example of a very large browser, the giraffe. Photo: Brian Wilsey.

sheep are mixed feeders. Mixed feeders are especially important in reducing woody plants in grassland systems due to their browsing activities. Long-term exclosures in Yellowstone National Park are full of willow and aspen, and these woody plants are reduced in abundance and stature when elk are present (Figure 2.8).

Many herbivorous mammals in grasslands are considered to be keystone species, as they form the basis of the **green food web**. The term keystone species can be vague and is sometimes used to denote any important species in an ecosystem. Some would argue that all species are important, so it is important to operationally define the term. **Keystone species** are species that have a larger effect on community structure and ecosystem processes than would be expected based on their relative abundance (p_i where p is abundance of a given species i/total abundance of all species). Elephants, which are considered to be keystone species, can completely uproot woody shrubs or trees to consume leaves, leaving savannas in a more open state. Wildebeests are keystone species in the Serengeti ecosystem due to their large numbers (~1.4 million animals) and migratory behavior (Sinclair and Arcese 1995), supporting huge predator and scavenger communities. For an example that demonstrates their importance to the whole ecosystem, a

Figure 2.8   Willows can grow to a height above the level of browsing when not browsed by elk in the northern range of Yellowstone National Park. Photo by Robert Beshta, Oregon State University—Tall willows with elk-winter, CC BY-SA 2.0, commons.wikimedia.org.

reduction of 50 percent in wildebeest from the 1.4 million Serengeti population would result in 700,000 fewer prey animals (Figure 2.9). Comparable numbers for a less abundant antelope species, the Coke's hartebeest (*Alcelaphus buselaphus*) with a population of 40,000 would be 20,000 less prey. Predators would be more strongly affected by a reduction in prey of 700,000 than a reduction of 20,000. This makes the wildebeest a keystone species due to its importance to predators and scavengers. The dung and urine produced by such a large population size can impact nutrient cycling (covered in more detail in Chapter 4).

In North America, bison are considered keystone species (Knapp et al. 1999). Bison have been shown to alter the amount of patchiness in North American grasslands with their patchy grazing, which alters fire intensity and movement, and ultimately, tree encroachment. They also affect nutrient cycling and litter build-up and can increase local biodiversity (Knapp et al. 1999). Grazed areas can have less fine fuel mass than non-grazed areas, which creates patches without fire or lower intensity fire (Knapp et al. 1999). Bison carcasses support a myriad of scavenger species (Wilmers et al. 2003), and the effects of these carcasses are hard to replace when bison are not present.

Another important factor in classifying herbivorous mammals is their digestive anatomy. Plant tissue is made up primarily of cell walls, which consist of

Figure 2.9   Wildebeest migration off of the Serengeti Plains into the savanna in Tanzania (Photo: Wikimedia commons). Wildebeest spend most of the rainy season in the plains and move off the plains at the beginning of the dry season.

cellulose and lignin. Mammal digestive systems lack enzymes to digest cellulose, so they are dependent on micro-organisms to do it for them. Cellulose is processed anaerobically (without O_2) in fermentation reactions, producing short-chain fatty acids and sugars that the herbivore uses as an energy source. Methane is an end product of metabolism, and it is estimated that ruminants are responsible for 20 percent of the global production of this important greenhouse gas, primarily from belching. (Interestingly, kangaroos produce less methane than ruminants do for some unknown reason.) The protein that herbivores get from their food is from the bacterial cells in the small intestine. As a result, most ruminants and other foregut fermenters obtain their nutrition primarily as microbial end products!

There are three distinct groups based on the ways that herbivores harbor cellulose degrading microbes: hindgut fermentation, foregut fermentation, and foregut fermentation with rumination (Figure 2.10). Monogastric animals have one stomach, and the microbes involved in fermentation of cellulose are found either near the stomach (**foregut fermenters**), or in the back of the gut in the cecum or large intestine (**hindgut fermenters**). Example of a foregut fermenters are the hippopotamus, kangaroos, and camels (including guanacos and llamas). Hippopotamuses, which evolved from cetaceans such as whales, dolphins, and porpoises, are major grazers in Africa that leave ponds at night to forage on grasses in the upland grasslands. Examples of hindgut fermenters are horses, zebras, and rhinoceroses. Hindgut fermenters can pass large amounts of food through their digestive system fairly rapidly, and as a result they can consume the lowest quality and highest quantity of food among the three groups. **Ruminants** are a specific type of foregut fermenter that have four-chambered stomachs (Figure 2.10) and have to chew their cud to preprocess the plant materials to help shred the material for the micro-organisms involved in processing the material. Food passes through the esophagus into the rumen, is then regurgitated and chewed again by the ruminant, and it then passes through the esophagus a second time. This processing of food is associated with having higher basal metabolic rates than non-ruminant foregut fermenters (Clauss et al. 2010).

As a result of these anatomical differences, the diets differ among these three groups, and so does their effects on the environment. Dung particle size is significantly different among groups, with smaller particles in ruminants than in other foregut fermenters or hindgut fermenters (Clauss et al. 2010). This difference in dung particle size can affect the types of dung beetles present as well as nutrient cycling rates. Ruminants consume a lower quantity but higher quality diet than hindgut fermenters, and include species such as cattle, bison, wildebeest, and white-tailed deer. The hindgut fermenter wild horses (zebras, asses, horses) consume the lowest-quality forage among the

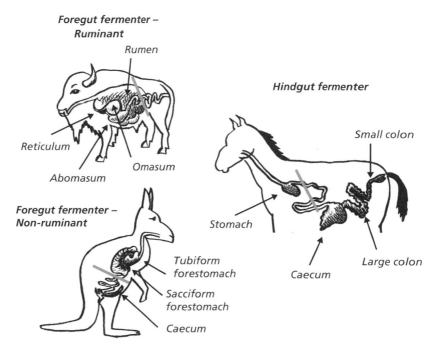

Figure 2.10 Digestive systems in herbivorous grassland mammals. Foregut fermenters harbor bacteria in the foregut region (note the gray line that denotes the junction between foreguts and hindguts), either in a rumen (ruminants denoted by a bison) or in a forestomach (denoted by a red kangaroo). Hindgut fermenters such as horses or zebras have 63 percent or more of their volume in the hindgut region. Foregut fermenters have a greater amount of volume in the foregut region. Digestive systems are not drawn to scale and are stretched out to ease the visualization.

three groups. Studies have found that feeding overlap is lower (niche partitioning is present) between feeding groups than within groups.

2.3.2 Coexistence of multiple large herbivorous mammals

These herbivore functional groups are useful for predicting how herbivores will impact each other when they co-occur. Mixed herds of wildebeest and zebra are common in Africa, and it was a long-standing mystery on how they coexisted together. Are they competing, or are they facilitating each other, or are they coexisting due to some other reason? Predation is not as important as a coexistence mechanism in large migratory species (Sinclair and Arcese 1995). Coexistence issues are also relevant when comparing how wild animals and livestock interact, and if they compete for food or not. Bell suggested that large herbivores in the Serengeti do not compete with each other, but rather facilitate each other due to the feeding types outlined previously, as well as differences in diet based on body size (Illius and Gordon 1992, covered in detail in Chapter 3). The hindgut fermenter zebra feeds on tall grasses and consumes a greater proportion of stems than other grazers (McNaughton 1985); wildebeest feed in patches with a greater proportion of leaves, and Thomson's gazelles feed on regrowth from wildebeest. This kind of niche partitioning allows coexistence by a large number (twenty-eight) of herbivorous mammal species in the Serengeti.

Many studies have assessed dietary-overlap in terms of food items consumed, usually using percent similarity measures (Table 2.1). These studies are typically done in species that eat similar things, so they tend to have higher overlap than they might with studies of species that are really different (e.g., carnivores vs. herbivores might have 0 percent overlap). This has to be taken into account when assessing overlap studies (Table 2.1). These studies are useful in assessing how the introduction of an exotic species will potentially affect native species, for comparing among and within guild species overlap, and for assessing how wild species might be impacting livestock and how livestock overlap with each other. Dietary similarity is higher between members within a feeding group (e.g., grazer vs. grazer) than between feeding groups (e.g., grazer vs. browser) (Table 2.1).

The types of herbivores feeding, and the number of herbivore species involved can be important. The effects of multiple herbivores feeding together vs. feeding by a single herbivore is understudied, but the studies that have been conducted suggest that there are different effects from multiple species feeding than from feeding by a single species. In these studies, it is important to compare grazing systems that have the same stocking rate or grazing intensity. Bagchi and Ritchie (2010) and Liu et al. (2015) did this. Bagchi and Ritchie compared herbivory by several native species (yaks, *Bos grunniens*; bharal, *Pseudois nayaur*; and ibex, *Capra sibirica*) and found 49 percent greater soil C storage than grazing by domestic species (cattle,

Table 2.1 Diet overlap measures between grassland animal species. Method denotes the percent overlap measure used.

Species pairs	Method	Overlap	Study, Country
Grazer vs. Grazer			
Horse (feral) *Equus caballus* vs. Blackbuck *Antilope cervicapra*	Schoeners	89	Baskaran et al. (2016), India
Swamp deer vs. Hog deer	PSI	92	Wegge et al. (2006), Nepal
Browser vs. Browser			
White-tailed deer *Odocoileus virginianus* vs. Black-tailed deer *Odocoileus odocoileus*	Horn's	92	Whitney et al. (2011). USA
Red deer *Cervus elaphus* vs. Apennine chamois *Rupicapra pyrenaica* (threatened)	Pianka	90	Lovari et al. (2014), Italy
Mixed feeder vs. Browser			
Elephant vs. Giraffe	Schoeners	80	O'Kane et al. (2013), Kenya
Elephant vs. Kudu	Schoeners	72	O'Kane et al. (2013), Kenya
Elephant vs. Nyala	Schoeners	81	O'Kane et al. (2013), Kenya
Elephant vs. Impala	Schoeners	73	O'Kane et al. (2013), Kenya
Giraffe vs. Kudu	Schoeners	76	O'Kane et al. (2013), Kenya
Giraffe vs. Nyala	Schoeners	77	O'Kane et al. (2013), Kenya
Giraffe vs. Impala	Schoeners	71	O'Kane et al. (2013), Kenya
Kudu vs. Nyala	Schoeners	78	O'Kane et al. (2013), Kenya
Kudu vs. Impala	Schoeners	80	O'Kane et al. (2013), Kenya
Nyala vs. impala	Schoeners	79	O'Kane et al. (2013), Kenya
Giraffe vs. Grant's gazelle	Oosting	30	Hansen et al. (1985), Kenya
Giraffe vs. Impala	Oosting	13	Hansen et al. (1985), Kenya
Giraffe vs. Elephant	Oosting	30	Hansen et al. (1985), Kenya
Elephant vs. Grant's gazelle	Oosting	42	Hansen et al. (1985), Kenya
Elephant vs. Impala	Oosting	49	Hansen et al. (1985), Kenya
Rhea vs. Plains viscacha, Spring	Kulczysnksi	33.8	Pareira et al. (2003)
Cattle vs. Rhea, Spring	Kulczysnksi	30.7	Pareira et al. (2003)
Mule deer vs. Sheep	Schoener's	31	Beck and Peek (2005), USA
Elk *Cervuus canadensis* vs. Sheep *Ovis aries*	Schoener's	40	Beck and Peek (2005), USA
Black tailed jackrabbits *Lepus californicus* vs. Cattle	Kulczysnksi	41	Daniel et al. (1993), USA
Grazer vs. Browser			
Cattle vs. Mule deer *Odocoileus hemionus*	Schoener's	6	Beck and Peek (2005), USA
Giraffe vs. Thomson's gazelle	Oosting	3	Hansen et al. (1985), Kenya
Giraffe vs. Topi	Oosting	1	Hansen et al. (1985), Kenya
Giraffe vs. Kongoni	Oosting	3	Hansen et al. (1985), Kenya
Giraffe vs. Wildebeest	Oosting	1	Hansen et al. (1985), Kenya

(continued)

Table 2.1 (Continued)

Species pairs	Method	Overlap	Study, Country
Giraffe vs. Zebra	Oosting	10	Hansen et al. (1985), Kenya
Giraffe vs. Buffalo	Oosting	2	Hansen et al. (1985), Kenya
Grazer vs. Mixed feeder			
One horned Rhinoceros vs. Swamp deer	PSI	70	Wegge et al. (2006), Nepal
One horned Rhinoceros vs. hog deer	PSI	70	Wegge et al. (2006), Nepal
Cattle vs. Plains Viscacha, Spring	Kulczysnksi	54.7	Pareira et al. (2003), Argentina
Cattle vs Elk *Cervus elaphus*	Schoener's	26	Beck and Peek (2005), USA
Prairie dogs vs. Cattle and goats	PSI	69	Mellado and Olivera (2008), Mexico
Przewalski's gazelle vs. Tibetan sheep	Schoeners	71	Liu and Jiang (2004), China
Predator vs. Predator (Carnivores)			
Feral cats *Felis catus* vs. an endemic Island fox *Urocyon littoralis*	Pianka	93	Phillips et al. (2007), USA
Bobcats *Lynx rufus* vs. Mountain Lion *Puma concolor*	Pianka	56	Hass (2009). USA
Common and snow leopards	Pianka	69	Lavori et al. (2013), Nepal
Spotted hyaenas *Crocuta crocuta* vs. Lion *Panthera leo*	Levins	59	
Wild dogs vs. Cheetahs	Pianka	74	Hayward and Kerley (2008), South Africa
Leopards vs. Cheetahs	Pianka	74	Hayward and Kerley (2008), South Africa
Wild dogs vs. Leopards	Pianka	74	Hayward and Kerley (2008), South Africa
Lions vs. Spotted hyaenas	Pianka	60	Hayward and Kerley (2008), South Africa
American badgers *Taxidea taxus* vs. Coyotes *Canis latrans*	Renkonnen	47	Azevedo et al. (2006), USA
American badgers *Taxidea taxus* vs. Red fox *Vulpes vulpes*	Renkonnen	55	Azevedo et al. (2006), USA
American badgers *Taxidea taxus* vs. Racoons *Procyon lotor*	Renkonnen	52	Azevedo et al. (2006), USA
American badgers *Taxidea taxus* Striped skunk *Mephitis mephitis*	Renkonnen	50	Azevedo et al. (2006), USA
Red fox *Vulpes vulpes* vs. Coyotes *Canis latrans*	Renkonnen	67	Azevedo et al. (2006), USA
Racoons *Procyon lotor* vs. Coyotes *Canis latrans*	Renkonnen	53	Azevedo et al. (2006), USA
Racoons *Procyon lotor* vs. Striped skunk *Mephitis mephitis*	Renkonnen	60	Azevedo et al. (2006), USA
Striped skunk *Mephitis mephitis* vs. Coyotes *Canis latrans*	Renkonnen	46	Azevedo et al. (2006), USA

Species pairs	Method	Overlap	Study, Country
Striped skunk *Mephitis mephitis* vs. Red fox *Vulpes vulpes*	Renkonnen	45	Azevedo et al. (2006), USA
Other			
14 species of grasshoppers	Kulczynski	4-91	Uekert and Hansen (1971)
Grasshopper species		35	Joern (1979), USA
Livestock vs. Livestock			
Cattle vs. sheep	Morisita	90	Schwartz and Ellis (1981)
Cattle vs. sheep	Kulcyznski	64	Celaya et al. (2007), Spain
Cattle vs. Sheep	Schoener's	56	Beck and Peek (2005), USA
Goats vs. sheep	Kulcyznski	63	Celaya et al. (2007), Spain
Cattle vs. Sheep	Qualitative	"Low overlap"	Falu et al. (2014), Argentina
Cattle vs. goats	Kulcyznski	50	Celaya et al. (2007), Spain
Cattle vs. sheep		83	Squires, USA
Cattle vs. goats		75	Squires, USA
Goats vs. Sheep		70	Squires, USA

yak-cattle hybrids, horses, donkeys, goats, and sheep) stocked at similar rates in India. Herbivory by domestic species led to about 25 g C m^{-2} year^{-1} less plant production aboveground compared to consumption by native herbivore species. Niche overlap in foods was lower in the native species assemblage than the domestic species assemblage. Liu et al. (2015) and Wang et al. (2010) compared single grazer vs. multiple grazer systems and found that grazer effects interacted strongly with plant species diversity in Chinese meadow systems. Increases in plant species richness due to grazing in diverse areas were greater when cattle and sheep were present compared to cattle or sheep alone. Wang et al. (2010) found that sheep prefer diverse diets over less diverse diets when given a choice, and that sheep consumption increased with plant diversity of the stands.

2.3.3 Small mammals

Small mammals are also abundant in grasslands. **Fossorial mammals** (latin *fossor*, or "digger") are animals that burrow underground and feed on roots. Pocket gophers and prairie dogs were once very common rodents in North American grasslands. Similar animals are cape ground squirrels in southern Africa, and plains *viscachas* in South America. These mammals are important in creating disturbed patches due to their burrowing and mound building. Pocket gophers continuously create disturbances around their mounds, and seedling establishment of rare forbs can be higher on mounds than away from mounds (Klaas et al. 1998). A myriad of mouse and vole species are usually present, which have important effects by feeding on seeds and seed-lings (Mortenson et al. 2017).

Prairie dogs, a North American fossorial species, have been described as a keystone species because of their effects on reducing shrubs (Weltzin et al. 1997) and engineering activities (**ecological engineers**) and because their colonies support endangered species. However, they have also been viewed as pests by ranchers, and the number of colonies or towns has declined by more than 90 percent from presettlement (Whicker and Detling 1988, Augustine and Baker 2013). Prairies dogs are most common in mixed grass prairie. Prairie dogs live in large towns and can be seen feeding between an extended series of holes that emerge from mounds. They have multiple bur-rows per animal and have an advanced social network. Animals sit upright to observe for danger, have elaborate warning calls, and will duck down when a dangerous animal is observed in the area. In addition to prairie dogs, the burrow system supports other species such as burrowing owls, which can often be seen in daytime standing near a hole, black-footed ferrets, and many reptiles and amphibians. Augustine and Baker (2013) found that on-colony sites had significantly greater densities of carnivorous birds: burrow-ing owls (*Athene cunicularia*), mountain plovers (*Charadrius montanus*), and killdeer (*Charadrius vociferus*), and omnivores consisting of horned larks (*Eremophila alpestris*), and McCown's longspurs (*Rhynchophanes*

Figure 2.11   Bison and prairie dogs (foreground). Prairie dogs elicit a warning when danger is in their area (foreground). Bison preferentially feed in prairie dog towns. Photo: public domain.

mccownii) than bird communities off-colony sites. Black-footed ferrets were close to extinction by the 1980s but have made a nice recovery through the reintroduction of individuals to prairie dog towns in protected areas. Bison and pronghorn coexist with prairie dogs (Figure 2.11), and bison have been shown to prefer grazing on prairie dog towns over non-dog sites (Detling 1998). Overall, prairie dogs greatly increase the spatial heterogeneity of mixed grass prairie by creating different conditions between on-colony and off-colony sites. Prairie dogs browse and kill shrubs such as mesquite, which kept prairies in a more open state in the past by preventing woody plant encroachment (Weltzin et al. 1997).

Grasslands also support an abundance of granivorous (seed eaters) and non-fossorial herbivores. Voles can be abundant, especially during peak years (populations fluctuate from year to year), and Kang et al. (2007) found that they can consume 15–44 percent of NPP during peak years. Granivorous rodents are especially important in prairie restorations, and Howe et al. (2002) suggested that the community that forms is basically the species that are not consumed by voles. Mortenson et al. (2017) found that rodent granivory can prevent plant outbreaks of an annual plant species *Chamaecrista fasciculata* in a Midwestern USA restoration.

2.3.4 Birds and other reptiles

Grasslands have a smaller number of bird species than forested systems, and the bird species that are endemic to grasslands tend to be specialized to open habitat. Many grassland birds are rare and in decline, due to habitat loss and grassland fragmentation. Grassland birds also tend to be ground nesters, and nests are more likely to be found by predators in linear and small habitats found in human modified environments. North American species that are less abundant than they used to be include the bobolink, dicksissel, the greater and lesser prairie chicken, and the meadowlark. In South America, grassland birds have declined in the *Pampas* and *Campos* grasslands due to intensification of agriculture (Azpiroz et al. 2012). In the European UK, Donald et al. (2006) found that 19 grassland species are in decline due to agricultural intensification.

Perhaps the most interesting grassland birds are the large flightless birds. Grasslands of the southern hemisphere support populations of large flightless birds, the ratites. The emu (*Dromaius novaehollandiae*) is a common bird in Australia in grasslands and savannahs. The ostrich and rhea are large flightless birds in African and South American grasslands, respectively. Kiwi are native to New Zealand. The extinct moa was especially large, reaching heights of over three meters in New Zealand. In some areas, ratites can be abundant and important native grazers. They have interesting reproductive biology, with nesting on the ground, paternal incubation of eggs, and in the case of ostriches, occasional non-parent incubation of eggs. Early taxonomic categories suggested that the ratites were all descended from a single flight-less bird ancestor and evolved into separate species as Gondwanaland continents drifted apart. However, recent molecular-based phylogenies have overturned this view, and suggest that the bird ancestors flew to their respective continents and then independently evolved a loss of flight through convergent evolution (Harshman et al. 2008). Rapid running to escape predators in open grassland environments may have been uniformly beneficial in these birds in different parts of the world.

Other grassland bird species, such as cattle egrets, have adapted to the growing cattle herds in grasslands around the world. Originally from Africa, they have spread to other continents are thriving by eating insects kicked up by cattle rather than by the antelope that they followed in their native range. Many grassland bird species evolved in association with large migratory herds of grazing mammals, and they have interesting modes of reproduction. An example is the brown-headed cowbird (*Molothrus ater*), native to the central and western USA. It originally followed herds of bison as they moved through prairies. Because cow birds were constantly on the move, they evolved nest parasitism—laying eggs in other bird species' nests rather than their own. These eggs hatch and their offspring are raised by the other bird species. As the East coast of the USA was cleared, the cow bird invaded formerly forested areas, where they have had negative effects on birds that are evolutionarily naive to egg parasites (Brittingham and Temple 1983).

Grasslands are also home of large herbivorous turtles (Figure 2.12). The leopard tortoise of Africa and the Galapagos tortoise of Ecuador are major herbivores. Leopard tortoises feed on forbs and grasses in East Africa and

Figure 2.12   Leopard tortoise in Tanzania, Africa. Leopard tortoises are grazers in this environment. Photo: Brian Wilsey.

primarily grasses in South Africa (Kabigumila 2001). Galapagos tortoise declines have led to reductions in ephemeral wetlands that are similar to the loss of bison wallows (Froyd et al. 2013). Predatory snakes are common in grasslands, and some are highly poisonous (e.g., rattlesnakes in North America and cobras in Africa and Asia). Lizards are also common in grasslands, especially arid grasslands (Pianka 1967).

2.3.5 Predatory mammals

Intact grasslands typically have large predatory mammals that preyed upon herbivores and detritivores. In Africa, there are a number of large predatory mammal species, with lions and hyenas being most important. Jackals are important scavengers, feeding on dead carcasses that are left by other mammals. New Zealand and Australia were without large placental mammal predators before human occupation. In Eurasia, wolves and bears were important top predators, as they were in North America. Top predators have been extirpated in many areas and reintroduced in some areas. Reintroduction of wolves has led to changes in feeding patterns by herbivorous mammals, which is having important top-down effects on vegetation. In Minnesota (USA), they are having important impacts on mesopredator abundances (medium sized species) due to their cascade effects on prey species: wolves reduce coyote numbers, which leads to increases in fox numbers, which have negative effects on birds (Levi and Wilmers 2012). Trophic cascades such as these will be detailed in Chapter 5.

2.3.6 Invertebrates

Insects and other invertebrates are common in grasslands, as they are in all ecosystems. Most forb species are animal pollinated, and their abundant pollen and nectar production supports a variety of bees, butterflies, and other insects. Flowers produce an abundance of pollen, which serves as a protein source and nectar for carbohydrates. A typical flower will receive visits from multiple species. Grasslands, especially diverse sites, are important to pollinating species such as bees and pollen feeders such as butterflies. Pollinators preferentially use diverse grassland sites, and these sites have been shown to be important to pollinators. In England, only four grassland plant species produce more than 50 percent of the nectar available to bees (Baude et al. 2016), and these species are supported by grassland conservation areas such as calcareous and neutral grasslands and woodlands. Pollinating insects are in decline in many parts of the world, and Biesmejer et al. (2006) found that out-crossing insect pollinated plants in Britain and the Netherlands are declining in concert.

Insects can be grouped into functional groups according to what they consume, with some **leaf feeders** feeding on leaves, **miners** that lay their eggs within the leaf (a type of parasite), and **sapsuckers** that feed on vascular juices. Sapsucking insects such as aphids and spittle bugs, tap into the plant's vascular system and can become locally abundant during outbreaks (Wilsey et al. 2002). Spittle bugs are especially interesting (Figure 2.13, also called frog-hoppers), with their spittle mass acting as a defense against predators. Inside the spittle, the bug taps into the plant's xylem, feeding on water and nutrients. **Detritivores** feed on dead plant tissues, either shed roots, leaves, or stems. Collembola (springtails) and crickets are important detritivores in grasslands world-wide. They both can become extremely abundant in certain years. Collembola are one of three hexapod lineages that are no longer considered to be insects. The detritivore food web, where organisms consume dead plant tissues and other forms of detritus, is sometimes called the **brown food web**. It is understudied compared to the green food web.

Many species of grasshoppers are found in grasslands, and they can be locally abundant, especially in lightly and moderately grazed areas. For example, there are 150 species of grasshoppers in Inner Mongolia Steppe, with 10–15 being considered to be pest species (Kang et al. 2007). Kang et al. found that there was strong niche partitioning in these species, and classified them into phytocarnivous, graminivorous, forbivorous, and mixed graminvorous-forbivorous based on their diet. There was also temporal partitioning with early, mid-year, and late in the year hatching species.

Figure 2.13   Spittle bugs (frog hoppers) consume xylem fluid, which is almost entirely water. Trace amounts of amino acids and minerals provide nutrition for the spittle bugs, and it is amazing that these insects can survive on such a low nutrient food supply. The spittle mass is an antipredator adaptation. Spittle bugs are fairly common in grasslands (Wilsey et al. 2002). Photo: Wikimedia.

Additional partitioning may come from grasshopper diets based on the proportion of protein and carbohydrates in their diets (Behmer and Joern 2008).
Grasshopper species richness peaks in the middle of the year and is highest
where plant diversity is highest in lightly grazed areas with intermediate biomass amounts (Kang et al. 2007). In tallgrass areas, grasshopper abundance
and species richness are higher in areas that are grazed by bison, which have
plant tissues that are higher in N, higher plant-species richness and diversity, and higher structural heterogeneity (Joern 2005).

2.4 Microbes

Single celled organisms are important in grassland nutrient cycling and as
mutualists and pathogens, and are extremely abundant in soil (van der
Heijden et al. 2008). The single celled organisms in the soil are collectively
known as the soil **microbiome**. Most microbes in grassland soils have never
been cultured and are known only from their DNA or RNA sequences. Next
generation sequencing approaches are opening up the world to the microbes
in grassland soils (Fierer et al. 2013). **Operational Taxonomic Units** (OTUs)
are used to delineate organisms that have yet to be described based on differences in their nucleotide base sequences, usually a 3 percent difference is
enough to classify organisms into separate OTUs. The domain Eubacteria
contain many species that are important in nutrient cycling, from nitrogen
fixation in legumes (*Rhizobia* spp.), to mineralization of organic matter to
inorganic NH_4 and NO_3 that is available to plants (ammonification, nitrification), to decomposition of chitin and cellulose. They also contain human
pathogens like the gut microbe *Escherichia coli*, which may enter water bodies where large mammals are present. Single celled fungi, which are
Eukaryotic, are important as decomposers in grassland soils, and commonly
grow on plants as fungal endophytes. The domain Archaeae, which were
originally thought to be found only in extreme environments, turn out to be
fairly common in grassland soils. Their function is largely unknown.

Bacterial diversity is interesting because it is not affected by some factors
that typically predict plant and animal diversity, such as temperature and
latitude (Fierer and Jackson 2006). Across a wide variety of systems, soil pH
was a strong predictor of bacterial diversity (as it is in plants as well), with
diversity higher in neutral than in acidic soils (Fierer and Jackson 2006). In
prairie grasslands of North America, precipitation was an important predictor
of bacterial diversity, and the phylum Verrucomicrobia dominated bacterial
communities across sites (Fierer et al. 2013). Plowing and conversion of
prairie grasslands to crop fields resulted in altered bacterial community
composition (Fierer et al. 2013).

2.5 Fungi and other decomposers

Due to the focus on grasslands as producers of animal products for humans, there has been fewer studies of taxonomic groups other than plants and animals in grasslands. Fungi are especially important as decomposers and as mycorrhizae. Mycorrhizae are fungi that grow on plant roots. Most plant species in grasslands have mycorrhizal fungal associations. In contrast to trees, which commonly harbor ectomycorrhizae on the outside of the roots, grassland species typically have arbuscular mycorrhizae, which are characterized by the fungus growing into root cells. There has been much research into this interaction, and typically the plant provides a carbon source to the fungus, and the fungal mycelia grow out from the plant into the surrounding soil. This is a mutually beneficial interaction (a mutualism) as the mycelia provides nutrients to the plant, especially P, and can help the plant with water uptake due to the greater surface area of roots plus mycelia. However, in times of stress, the plant can cut-off the carbon supply to the fungus, and in some situations, mycorrhizae can become parasitic on the plant. Klironomos (2003) suggests that mycorrhizae are not always mutualists in grasslands, with interactions ranging from +/+ to +/0. Plant species introduced by humans had smaller positive growth responses to mycorrhizae than native plant species (Pringle et al. 2009). Mycorrhizal species can also switch among plant hosts, which helps them from becoming 'enslaved' by the plant.

Perhaps most interestingly, mycorrhizae can develop connections between separate plant species in at least some cases. We know this because Grime et al. (1987) labelled plants with an isotopic label and found that C was being transferred between plant species. This transfer was missing when mycorrhizal connections were severed. There is still much to be learned about plant-fungal interactions, and especially about interspecies resource transfer. Other major decomposers in grasslands are bacteria and various detritivorous invertebrates, which will be covered in Chapter 4.

2.6 Animal migrations

The large animal migrations that still occur in grassland systems are important reminders of how sedentary animal grazing systems will not easily replace the original (Figure 2.9). Grazing mammals were mostly migratory to varying degrees in the past, most commonly following spring green-up along elevational gradients (Frank and McNaughton 1992), and following green-up with dry to wet season transitions in the tropics (McNaughton 1985), or seasonal movement to calving grounds. Migrations are no longer possible in many parts of our current human-impacted Earth, but the migrations have been studied extensively in the Serengeti of Tanzania and Kenya, and in

Yellowstone National Park. Migrations of the Tibetan antelope (*Pantholops hodgsonii*) have also been studied (Xia et al. 2007) and were the focus of a Motion Picture documentary. Migrations of wildebeest and zebra in the Serengeti follow rainfall patterns, and animals leave the short-grass plains at the end of the rainy season. They move through mixed grass transitional areas until they reach tall-grass areas where they spend the dry season months. They again move back to the short-grass plains at the beginning of the next rainy season. They have been found to optimize their nutrient uptake over the year by moving between these different areas (McNaughton 1988). They also move around extensively within the short-grass plains during the wet season. In the Yellowstone system, bison migrate up in elevation during spring and early summer following spring green-up. They spend summer months at high elevations before returning down in the late summer, fall. This movement has been linked to greater forage quality in their diets compared to staying in one place (Frank and McNaughton 1992). Importantly, these movements prevent areas from being grazed repeatedly in a single year, giving them adequate time to recover (Oesterheld and McNaughton 1992). These migratory patterns should be understood by managers of cattle and other livestock. Migratory patterns are mimicked in livestock systems by rotating animals among feeding paddocks or by adjusting stocking rates to allow grassland rest between grazing events.

The migration of grazing mammals can have important effects on predatory animals. Predators can either follow the herds, as do spotted hyenas in the Serengeti, or they can stay in place and let the herds pass through their territories, as lions do in the Serengeti. Female hyenas that are raising young have been observed to travel very long distances to and from the herds to bring food back to their young. This involves large costs to their fitness due to travel. Lions remain in a territory and encounter 'boom or bust' conditions depending on whether the herds are within their territory (boom) or not (bust). Sinclair and Arcese (1995) found that this has important effects on whether grazing mammal populations are regulated by predation or not. In migratory populations, grazers were not limited by predators but by food and accidents that occur during migrations. In the non-migratory populations in the Ngorongoro crater, antelope were regulated by predators.

An important fact that is often overlooked is that not all individuals of migratory species are migratory even in areas that still have migrations. Non-migratory individuals of a wildebeest and zebra have been studied in Africa. They have important phenotypic differences compared to migratory individuals, including larger body mass and shorter legs. Non-migratory individuals are much less abundant than migratory individuals in these areas, suggesting that it might be beneficial to migrate. Resident herbivores can have important effects on nutrient cycling, sometimes increasing nutrient availability and in turn, increasing their own carrying capacity (McNaughton et al. 1997).

3 Factors Maintaining and Regulating Grassland Structure and Function

3.1 Disturbance

Grasslands evolved with fire or herbivory, or both, which are important disturbances. **Disturbance** is defined as any factor that damages biomass (Grime 1977). There are two kinds of disturbances, extrinsic or intrinsic, and they have different effects. **Extrinsic disturbances** are those that came from outside the normal experiences of the organisms present, including human activities like plowing, removal of topsoil during mining activities, and vehicle tracks. **Degradation** is sometimes used to describe human-caused extrinsic disturbances. Plowing is the most common disturbance affecting grasslands globally, and it has been used to transform native grasslands into crop fields and simplified pasture. Recovery from plowing takes a long time and is possible only when the site is near extensive non-plowed areas that provide a propagule source (Bakker and Berendse 2002). Extrinsic disturbances were not common in the evolutionary history of grasslands, although they sometimes mimic disturbances that were (e.g., trampling from vehicles may be similar to trampling from large mammals). **Intrinsic disturbances** are processes that have occurred during and after grassland formation on an evolutionary time scale. Examples of intrinsic disturbances include fire, wind-damage, digging or burrowing by fossorial mammals, defoliation, and trampling by native large mammals. Grassland species evolved with intrinsic disturbances, and they can be important in maintaining grassland community structure and functioning. Removal of intrinsic disturbances can lead to degraded grasslands and tree encroachment (see Chapter 8). Responses to intrinsic disturbances can sometimes be different from responses to extrinsic disturbances. Grasslands can be irrevocably altered by extrinsic disturbances such as plowing but can be highly resilient to intrinsic disturbances. Thus, the impact of extrinsic disturbances partially depends on how different they are from intrinsic disturbances.

The Biology of Grasslands. Brian J. Wilsey. Published 2018 by Oxford University Press. © Brian J. Wilsey 2018.
DOI 10.1093/oso/9780198744511.001.0001

3.2 Fire

Fire is an important intrinsic disturbance in grasslands. Grassland litter (senesced biomass) is highly flammable, and over several years without fire or grazing, fuel will accumulate due to grassland's relatively low decomposition rates. Combined with occasional droughts and frequent lightning storms fires are inevitable. Fire represents an alternative pathway for biomass if biomass is not removed by grazing (Bond and Keeley 2005), and it is especially important at the high precipitation end of the grassland spectrum, where fuel is more abundant.

Fire dynamics have changed over time due to human activities. Humans have altered the amount of area burned and the time intervals between fires. Changes in fire combined with human-caused increases in seed pressure from trees, have favored a tree-dominated state. Fires that would have burned widely in the past are now extinguished by humans in most parts of the world. Moreover, in many regions, the highly fragmented nature of modern landscapes greatly hampers the spread of fires once they start. A fire that would have spread widely across a huge portion of the landscape in former times will stop now when it gets to the edge of a field or road, or some other human modified object. This is in contrast to the past activities by some cultures (e.g., native American groups), who's members frequently set fires in a more intact system, which kept grasslands open even in areas that would support forest without fire (Nowacki and Abrams 2008). Furthermore, human planting and protection of trees has resulted in areas that may have received few seeds in the past now being under constant pressure from tree and shrub seeds. Under high propagule pressure, the infrequent fire cycle that may have been sufficient in maintaining grassland in the past may be insufficient now. Managers sometimes counter increased pressure from woody plants with a more frequent fire cycle than we would have seen in the past, this change in fire frequencies also may have negative ramifications for grassland organisms.

3.2.1 Fire intensity

Plant species vary widely in their tolerance of fire. Most woody plants are killed by fire if they are small in stature. Some savanna tree species will survive fire if they are tall enough to escape the fire; once they escape the fire they typically survive future fires due to their thick fire-tolerant bark (Bond and Keeley 2005). Browsing mammals like giraffes can interact with fire to keep seedlings within the fire zone. Members of the *Quercus* and *Acacia* genera are well-known for their tolerance of fire once they become adult trees. Grasses and most grassland forbs are tolerant to fire, grasses due to their underground crowns (Figure 3.1) and their herbaceous life forms.

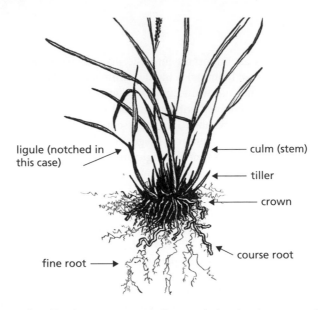

ligule (notched in this case)

culm (stem)

tiller

crown

course root

fine root

Figure 3.1   Typical grass plant (*Sorghastrum nutans*, indian grass), showing the crown, which is the base of the plant and the underground perennial storage organ. Shoots and tillers extend from the crown. The ligule, which is usually at the base of the leaf, contains anatomical features used in identification. Only the outside part of the plant is shown.

The ability to die down to underground organs, exposing only dead tops aboveground, is a key feature of grassland plants (Gleason 1922).

Three quantifiable measures of fires are: their 1) **frequency** (how many fires occur over a given time period), 2) **intensity** (how hot they are when they occur), and 3) **timing** (e.g., spring, summer, fall or winter). These measures can predict whether grasslands will remain unforested, whether native grasslands resist invasion by exotic species, and whether early or later flowering species will be favored by the fire.

Fire frequency is correlated with fuel (litter) buildup, so it is more frequent in areas that receive higher precipitation. In shortgrass regions, and in areas that have higher grazing pressure, fire can be less frequent because of low litter accumulation rates. In these systems, the fire frequency may be measured in decades. Subhumid grasslands like tallgrass prairie typically have a fire frequency of five years or less, which means that any given site would be burned every five years. Two long-term studies have given us valuable information on fire frequency: Konza Prairie in Kansas, USA and Krueger National Park in South Africa (Box 3.1). A long-term study of fire frequency demonstrated that the highest plant-species diversity was found at fire frequencies (two to four years) that match the natural frequency of the area (Blair 1997). Annual burning, which is common on ranches used to raise cattle, leads to high dominance by C_4 grasses (Collins et al. 1998).

Box 3.1 Konza prairie research area

Konza Prairie has multiple watersheds that receive fire in the spring every year (1-year cycle), two years (2-year cycle), four years (4-year cycle), or 30 years (30-year cycle). Half of the watersheds are grazed by bison (*Bison bison*) to examine grazer interactions with fire frequency. Major conclusions of this study are: 1) shrub and tree encroachment are lowest in the annually burned areas, 2) plant diversity is highest with a four-year fire frequency, 3) bison grazing affects fire intensity and reduces the decline in plant diversity that occurs with annual burning (photo from Wiki Commons).

Fire intensity can be estimated by measuring the proportion of fuel that was combusted, the color of the ash, or by measuring the temperature of the fire with thermocouples or meltable paints (Stronach and McNaughton 1989, Martin and Wilsey 2012). Intensity can vary widely across grasslands, and is higher when fuel loads are high, plants are dormant and present as standing dead biomass and litter, and in the presence of certain fire-promoting species. Fire-promoting species tend to be dormant and dry when the fire passes through, and fire-suppressing species tend to be green and wet when the fire passes through. Fire temperatures can vary from 83–680°C (Wright 1974), and the effects of fires on recovery will vary tremendously across these temperatures.

The species composition of plants, especially of non-native species can alter fire intensity. Invasive grasses such as cheatgrass (*Bromus tectorum*), which is an annual species, have been shown to increase the intensity and frequency of fires in semiarid steppe of the intermountain west USA (D'Antonio and Vitousek 1992, Balch et al. 2013). Grasses invading shrublands of Hawaii can cause fires in areas that normally do not burn, and the grasses, once established, make future fires even more likely (D'Antonio and Vitousek 1992). Semi-arid grasslands and some deserts which typically had a century-scale fire frequency now burn more frequently (Balch et al. 2013). However, this positive feedback is not found everywhere, and in cases where C_3 grasses are the invaders into mixed grassland, the feedback can be negative. Invasive grasses that are perennial and that are actively growing and green during the fire, such as *Bromus inermis* and *Festuca arundinacea* have been shown to reduce spring fire temperatures in more subhumid grasslands of the Midwestern USA (Martin and Wilsey 2012, McGranahan et al. 2012). Thus, these grasses have fairly general effects on altering fire intensities, but the effect can be positive or negative depending on whether the invasive grass is dormant during fire (heating effect) or mostly green during the fire (cooling effect).

In shortgrass steppes, fire was originally viewed by researchers as being infrequent due to lower litter masses for fuel, or of having negative effects due to their disturbance effects on biological crusts. However, the role of fire in shortgrass areas has been intensively studied recently, and some of our views have been modified. A study done by Augustine et al. (2010) that experimentally burned shortgrass steppe in March found that forage quality and N availability to herbivores were higher in the first summer after burning, but that there was no effect on net primary productivity. Plots fully recovered to preburn levels by the second growing season.

Fires can be harmful in steppe if they harm biological crusts. Biological crusts are soil aggregates at the surface of the soil held together by algae, fungi, cyanobacteria, lichens, and mosses (Johansen 1993, Liu et al. 2009). They are important because they harbor N fixing cyanobacteria and lichens, and they can contribute large amounts of N and organic C to soil. They stabilize soil surfaces, reducing runoff following rainfall, and can greatly affect seedling establishment of both native and exotic plant species. However, they are fragile and can be damaged by hooves from grazing mammals and fire in the wrong season. Ford and Johnson (2006) found that dormant season fire reduced N fixation and chlorophyll production in biological crusts in steppe. Liu et al. (2009) found that high grazing intensity lowered N input from crusts in Asian steppe.

3.2.2 Fire timing

The timing of fire can also lead to very important outcomes for a given fire event. Although most records of wildfires show that fire can occur anytime of the year, they are most common in hot and dry periods in subhumid areas (Nelson 1985, Howe 1994), and at the beginning of the dry season in tropical areas (Stronach and McNaughton 1989). This contrasts with the typical time periods for human prescribed fires, which are most often set in the winter or spring (dormant seasons, Howe 1994). Howe (2011) found that summer fires had very different effects on prairies than spring fires: summer fires favored early flowering forb species and spring burning favored C_4 grass species. Strong dominance by C_4 grasses from repeated spring burning can reduce species diversity (Collins et al. 1998). For this reason, prescribed fires (usually set in the spring) may not mimic the full effect of wildfires. Burning areas during different seasons could lead to patchiness in species dominance or diversity.

3.2.3 Plant adaptations to fire

Adaptations to fire include crowns and basal meristems in grasses (Figure 3.1), short stature, high allocation to belowground plant parts, ability to resprout after fire and herbaceous growth form. Most of these traits also appear to be adaptations to grazing and drought (Coughenour 1985), which suggest a syndrome in responding to the grassland environment. Trees have a large proportion of their biomass in aboveground parts that can be consumed by fire; herbaceous species do not.

One interesting adaptation that is less-frequently studied is the ability of seeds to respond to smoke with increased germination (Jefferson et al. 2008). Smoke has been found to increase seed germination in some Mediterranean shrubland species in Australia (Dixon et al. 1995). Nearly half of the species tested by Dixon et al. (1995) had increased germination when exposed to smoke from fire. Smoke stimulation of seed germination has been tested less often in grassland species, but there is evidence that seed germination of the common African grass red oat grass (*Themeda triandra*) is stimulated by smoke (Baxter et al. 1994). Schwilk and Zavala (2012) found that four out of fifteen grassland species tested showed smoke stimulation. Positive germination increases due to smoke might be important in restorations and other situations where recruitment is from seed and not vegetative propagules.

3.2.4 Fire and grazing interactions

Fire can greatly interact with the activities of grazing mammals because grazing reduces litter (fuel) load, and fires affect forage quality. First, intensively grazed areas rarely burn naturally. One of the major tests of how grazing mammals affect fire frequency occurred when wildebeest in the Serengeti recovered from rinderpest. The shortgrass plains in the Serengeti burned when wildebeest populations were low due to a rinderpest outbreak. Vaccinating cattle in areas surrounding the park reduced rinderpest, and

Figure 3.2   The green flush that is found when a grassland is recovering from fire is highly nutritious to herbivores. Notice the lack of standing dead biomass or litter on the soil surface. These areas can attract herbivores during the early regrowth period. Photo: Brian Wilsey.

wildebeest populations increased from around 250,000 in the 1960s and 1970s to around 1.3 million in the 1990s (Sinclair and Arcese 1995). As the wildebeest populations recovered, a smaller and smaller proportion of the shortgrass plains burned. Grazing represents an alternative to fire for litter consumption (Bond and Keeley 2005), and heavy localized grazing can lead to having unburned patches within a burned field (Knapp et al. 1999). Second, burned areas recover fairly quickly once rains return, turning the burned area into a **green flush area** that attracts herbivores (McNaughton 1985, Wilsey 1996, Sensenig et al. 2010, Figure 3.2). Plant tissues in the green flush are higher in water and nutrient content than unburned areas, and net primary productivity is higher during the regrowing period (McNaughton 1985). Small-body mass species, and foregut fermenters show stronger preference for green flushes than do large-body mass species and hindgut fermenters like zebra (Wilsey 1996, Sensenig et al. 2010, Figure 3.3). Reduced plant heights can also reduce predation risk to herbivores, although Eby and Ritchie (2013) found no evidence of increased predator vigilance in burned areas that had a cheetah decoy present. As a result of interactions between fire and grazing, concurrent grazing and fire lead to a patchwork of areas with different intensities of grazing and fire.

Preference for post-burn green flushes is higher in small bodied grazing mammals, and declines with body mass (Wilsey 1996). Preference is also related to herbivore functional group in that preference is higher in small

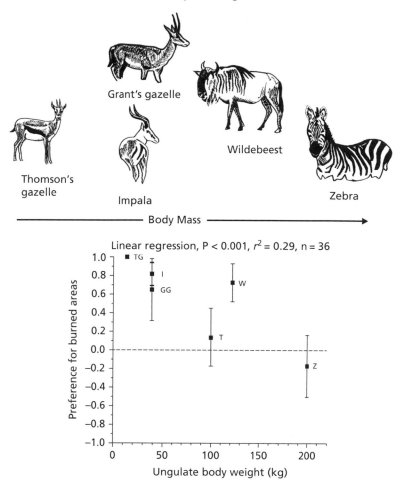

Figure 3.3 Preference for green flushes following burning in three paired burned and unburned grasslands (observed vs. expected). A gravel road served as a fire break between burned and unburned areas. Preference declined with mass. Plot of preference and body weight from Wilsey (1996), used with permission.

foregut fermenters (ruminant antelope) than it is in large hindgut fermenters like zebra (Sensenig et al. 2010).

Patchiness resulting from fire and grazing has been promoted by Fuhlendorf and Engle (2001) as a way to increase the amount of spatial variability in grasslands, the **patch-burn grazing model**. With patch-burn grazing, a portion of each field is burned each year, and grazing mammals preferentially graze the green flush that occurs after burning. By burning only a portion of the field each year, and because grazing mammals move onto newly formed

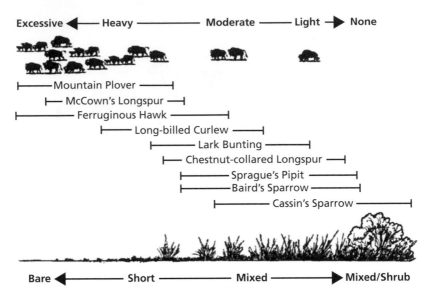

Figure 3.4   Bird species prefer different vegetation heights as habitat. Having a different level of grazing or fire intensity can be used to create the variation in heights (Figure from Derner et al. 2009, Reprinted from *Rangeland Ecology and Management*, Vol 62, pages 111–18, Derner et al. 2009. Livestock as Ecosystem Engineers for Grassland Bird Habitat in the Western Great Plains of North America, used with permission).

green flushes after the burn, the manager can prevent any one area from receiving heavy chronic grazing. Large variation in plant heights develops over time among patches, with shorter, recently burned areas coexisting with taller unburned areas nearby (Fuhlendorf et al. 2006, Figure 3.4). This has been shown to result in improved bird habitat for a great variety of species (Fuhlendorf et al. 2006). Henslow's sparrows, a species of concern, were found more frequently in the burned/grazed patches and less frequently outside patches. Upland sandpiper and a few other species were found more in recently burned-grazed patches. Thus, a greater diversity of grassland bird species is possible with patch burn-grazing treatments compared to homogenous grazing and fire regimes.

Shifts in species composition that occur after fires and grazing events can indirectly affect forage quality. These **indirect effects**, defined as effects that operate through a third species or functional group, can sometimes be of greater strength than direct effects. Anderson et al. (2007) found that fires in Africa favored the grass species red oat grass *Themeda triandra*, which has low forage quality. Over time, and after repeated burnings, the dominance of this species had a reducing effect on forage quality overall. Repeated spring burnings at Konza favor C_4 grasses (Collins et al. 1998), which have lower forage quality than C_3 species (Wilsey et al. 1997). This shift in functional group composition can indirectly reduce forage quality (Figure 3.5).

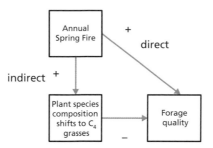

Direct vs. indirect effects in ecology, at Konza for bison
- Greater C_4 grasses, less legumes and C_3
- Greater cover of woody vegetation

Figure 3.5   Fires can have short-term positive direct effects on forage quality for herbivores. However, longer-term indirect effects can occur if fires favor C_4 grasses over C_3 species. C_4 grasses have lower forage quality than C_3 species. Indirect effects can be large and are important in understanding an ecological disturbance.

Having strong native species dominance is important to the patch-burn grazing model, in that it works best in prairie dominated by native plant species that are functionally dissimilar. Patch-burn grazing in fields dominated by the exotic plant species tall fescue were less successful at creating the spatial heterogeneity seen in studies with native prairie (McGranahan et al. 2012). Plant and animal diversity declined as the proportion of tall fescue increased independently of the patch-burn grazing patterns. This example points out the importance of non-native species, which will be covered in more detail in Chapter 7.

3.3 Herbivory

Herbivory is common in grasslands, and the origin of grasslands corresponds with the rise of large grazing mammals in most parts of the world (Edwards et al. 2010). Some intact grasslands receive rates of herbivory that rival some aquatic systems (McNaughton et al. 1989). The effects of herbivory on grassland plants depend on the type of herbivore (grazer, browser, mixed feeder), the types of plants being consumed (grass vs. forb), the evolutionary history of grazing in the area, and perhaps most importantly, the intensity (frequency and amount) of herbivory.

Is herbivory harmful to grassland plants and ecosystems? Herbivory has been extensively studied in grasslands, and attitudes changed during the twentieth century from the view that plant-herbivore interactions are a strict predator-prey (+ −) interaction, to the view that plants can tolerate some level of herbivory (+ 0) in many cases, especially in grasslands. In the 1980s, there were great animated debates about whether grazing was always detrimental to plants, and the two sides of the debate often used different measures of impact to support their claims (Belsky 1986, McNaughton 1986, Milchunas et al. 1988). The effect of herbivory can be quantified by measuring evolutionary fitness in plants. The effects of herbivores on plant fitness across all ecosystems found that it has a net negative effect on average (Hawkes and Sullivan 2001). However, plant fitness is more difficult to quantify than one might think in grasslands because most species are clonal and reproduce vegetatively (Figure 3.6). Vegetative propagules such as tillers are not considered in most studies of plant fitness; estimates are made solely on seed production. Seed production typically declines with grazing. However, when vegetative propagules are included in calculations, it is found that the number of new tillers produced can be higher with moderate grazing. Grazing tends to increase vegetative reproduction, and this increased tillering can lead to high plant tolerance to grazing. If fitness is based solely on seed production, then grazing is negative (Hawkes and Sullivan 2001); if it is based on seed and vegetative propagule production, then it can range from negative to neutral, and in a few cases to positive. Most of the positive cases are found in grassland systems.

One important thing to keep in mind when considering the effects of grazing is that there is great variation among grasslands in tolerance and evolutionary history of grazing (Figure 3.6). Milchunas and Lauenroth (1993) suggested that grasslands with a long history of heavy grazing will tend to respond

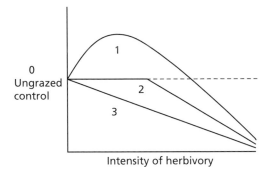

Figure 3.6   Responses to herbivory, usually measured as how plant growth or net primary productivity (y axis) varies with the intensity of herbivory (x axis), and between some grasslands that have evolved with heavy levels of herbivory and other non-grassland systems such as deserts that did not evolve with herbivory. The herbivore optimization model (McNaughton 1979, Hobbs 1996, denoted with a '1') posits that growth and productivity will be higher than controls (dotted line) with low to moderate levels of herbivory, and lower than controls with high levels of herbivory. Tolerance models posit that herbivory will have no significant effect on growth until herbivores over-ride the tolerance mechanism (denoted with a '2'). In some systems, herbivory causes a decline in plant growth at all levels of herbivory ('3'). The herbivore optimization and tolerance curves are especially common in grassland systems. Figure redrawn from McNaughton (1979).

differently to grazing compared to areas with little or no history of grazing. Examples of areas with a long history of grazing would be the short and mixed grass prairies of the central USA and the Serengeti, and examples of area without a history of grazing would be the Great Basin grasslands between the Rocky Mountains and the coastal ranges in North America, New Zealand, and the Pampas of South America. Bison were abundant in the central USA but were supposedly not found west of the Rocky Mountains, according to native American reports to the explorers Lewis and Clark. Great Basin grasslands are predicted to be less able to recover from grazing and should have greater species compositional shifts compared to grasslands east of the Rockies. Grasslands that have a long evolutionary history of grazing tend to have species that are either resistant or tolerant to herbivory (Strauss and Agrawal 1999, Milchunas and Lauenroth 1993). **Resistance** to herbivory is defined by having traits that minimize the amount of tissue removed by herbivores. **Tolerance** to herbivory is defined by traits that enable plants to recover from herbivory (Strauss and Agrawal 1999). Either resistance or tolerance can reduce or eliminate the reduction in fitness due to herbivory.

A concept related to tolerance is **compensatory growth**. The amount of tissue lost to herbivores rarely translates into the same amount of a biomass reduction or fitness. For example, if an herbivore removes 50 percent of the tissues of a plant, fitness declines are often less than 50 percent, sometimes considerably so. The reduction from expected is due to compensation by the plant. Overcompensation occurs when growth or biomass production is higher with herbivory than without (Figure 3.6). Overcompensation has rarely been found, but it has been found in a sedge from the shortgrass Plains of the Serengeti (McNaughton 1979), and in about 17 percent of the studies reviewed by Milchunas and Lauenroth (1993). Biomass production (biomass at the end of the experiment + biomass removed by grazers) was higher in grazed situations than in non-grazed situations. Undercompensation is found when the amount of biomass production is less than in ungrazed situations. Compensation is of obvious importance in managing grazing regimes. Stocking rate can be increased in areas with tolerant species that show large amounts of compensatory growth and should be kept low in areas where undercompensation occurs.

3.3.1 Adaptations to herbivory

Adaptations that enable grassland plants to resist grazing are morphological and chemical and are similar to plant adaptations to fire. Morphological adaptations include having crowns and basal (intercalary) meristems in grasses, short stature, high allocation to belowground plant parts, and herbaceous growth form. Being short is important in resistance to grazing because the height of grazing tends to be set by the herbivore mouth morphology. Defoliating a plant to 2 cm removes 50 percent of the foliage of a 4-cm tall plant, but 98 percent of the foliage of a 100-cm tall plant. The effect of 50 percent tissue loss will be less detrimental than 98 percent all other things being equal. The same logic explains how allocation to roots enables plants to resist grazing. Removing half of the biomass aboveground will have a more significant impact on a plant that allocates 80 percent of its biomass aboveground (40 percent lost), than one that allocates 20 percent of its biomass aboveground (10 percent lost). Herbaceous growth form is also important, as a plant has to replace its aboveground plant parts every year anyway in temperate areas and tropical areas with a dry season.

A few grassland species contain toxic compounds, or other chemical or mechanical defenses that help with resistance to herbivory. Low protein or N content (Table 3.1), especially in plants with C_4 photosynthesis, is considered to be an adaptation to grazing because of a common nutrient limitation in herbivores. Plant tissue is low in most nutrients and is difficult to digest. Cellulose and hemicellulose can only be digested by microbes in animal guts. High cellulose, hemicellulose, or silica concentrations, and low nutrient concentrations can lower digestibility to critical levels (van Soest

Table 3.1 Crude protein and N contents of various foods. Crude protein is N x 6.25, so N and crude protein are essentially the same. From Robbins, C. T. (1993). *Wildlife Feeding and Nutrition*. London and New York: Academic Press.

Food item	Crude protein (%)	Total (g)
Early growth in grasses, forbs, and browses	20–30	
Mature tissue in grasses, forbs, and browses	3–4	
Seeds	9.5–33.8	
Corn	8–15	
Fruits (80–90% water)	2–10 dry weight	
Fish, birds, mammal tissues	mean 60, range 22–91	
Invertebrate tissues	mean 65, range 30–88	
Legumes vs. Non-Legumes	Legumes > Non-legumes	
C_3 plants vs. C_4 plants	$C_3 > C_4$	
Example diets (30 g consumed):		
Mature plant tissue (herbivore)		
Species a (10 g)	3	0.3
Species b (20 g)	4	0.8
Total (protein limited)		1.1
Vertebrate animal tissue (carnivore)		
Species a (10 g)	6	6
Species b (20 g)	12	12
Total (energy limited)		18 g

1982). Silica can be abundant in grass tissues, and it is used as a defense against many insect and small mammal herbivores (Massey and Hartley 2009). Large mammals have molars that can crush grass tissues high in silica and can tolerate high levels. Grasses have relatively few species with common C based secondary compounds compared to forbs (Coughenour 1985).

Forb species are more chemically complex than grasses and sometimes have resistance mechanisms. Some forb species have toxins that make them poisonous to grazers. Milkweed species (*Asclepias* spp.) famously have cardiac glycosides and some alkaloids that make them unpalatable to all but a few host specific species such as the monarch butterfly (*Danaus plexippus*). In cases where selective herbivores are present, plant species that are resistant (unpalatable) can increase in abundance (Augustine and McNaughton 1998). An interesting question that awaits further answers is whether resistance and tolerance are positively or negatively correlated; tradeoff theory would predict that resistant plants would have lower tolerance to herbivory. Thus, resistance may not be a viable strategy if the resistant species are eventually defoliated (Augustine and McNaughton 1998).

3.3.2 Associational defenses

When palatable and unpalatable (defended) plant species coexist, it can lead to an interesting form of defense called **Associational Defense**. Associational defense is found when rates of herbivory are lower in a palatable species due to the presence nearby of a heavily defended species (Oesterheld and Oyarzábal 2004). Palatable plant species take advantage of the presence of unpalatable species and have higher survival and reproduction under heavily defended plants. For example, some grasses in African grasslands have higher growth and flowering rates when they are found under *Acacia* shrubs, which have thorns as an anti-herbivore adaptation (McNaughton and Tarrants 1983). A second famous example is a buttercup and grass association. Atsatt and O'Dowd (1976) reported on a study that removed the unpalatable and noxious buttercup species *Ranunculus bulbosus*, which contains the skin irritant ranunculin (a lactone). The grasses *Agrostis* and *Festuca* benefitted from growing under the *Ranunculus* species. Grass grazing by cattle in plots where *Ranunculus* was present was lower than in plots where it had been removed. Associational defenses are examples of when the community context of interactions matters a great deal.

3.3.3 Structural and chemical defenses

Defenses against herbivory can be structural or chemical, and chemical defenses can be inducible or constitutive. Forbs can have a variety of secondary compounds that act as defenses against herbivory. **Structural defenses** include thorns in some legumes and cacti, and **chemical defenses** such as

C-based phenolics or N-based toxins like glucosinolates in mustards and cyanide in some legumes. Defenses can be **induced**, which means they increase after herbivory, or can be **constitutive**, or produced with or without herbivory (Haukioja 1991).

Although forbs have higher concentrations of defenses than graminoids, they do not all respond with resistance; sometimes they tolerate and even overcompensate for herbivory. Paige and Whitham (1987) found that elk browsing of scarlet *Gilia* flowering shoots resulted in multiple shoots being produced in regrowth, thus increasing fitness over unbrowsed individuals. This was highly dependent on having most of the shoot removed, and the overcompensation effect was found to be highly conditional.

3.3.4 Compensation, tolerance, and grazing lawns

Compensation is especially common in graminoids of grassland systems (Table 3.2), and it can lead to plant tolerance to herbivory. The mechanism behind tolerance has been found to be associated with shifts in allocation towards leaves, release from apical dominance and the production of axillary shoots, and sometimes, shifts from roots and crowns to replace aboveground tissues. Shifting allocation from roots to shoots can have important implications on soil C storage. Plants can shift allocation from roots to shoots to replace lost tissue, can shift allocation from stems to leaves (Anderson et al. 2007), can increase leaf level photosynthesis, and can have higher growth rates due to higher leaf N concentrations that are found after herbivore events. Plant traits related to compensatory growth include having higher N concentrations

Table 3.2 Documented evidence for each mechanism contributing to compensatory growth following defoliation (from Hobbs 1996, used with permission).

Level of organization	Mechanism
Individual plant	Increased allocation of photosynthate to leaves
	Increased photosynthetic rate
	Increased uptake of nutrients by roots
	Increased allocation of nitrogen to leaves
Plant community	Reduced transpiration
	Increased water use efficiency
	Increased light penetration
	Increased soil moisture
Ecosystem	Increased mineralization of nitrogen
	Reduced immobilization of nitrogen
	Rapid return of mineral nitrogen to plants via excretion of urine and dung

and photosynthetic rates in regrowing foliage, having less water stress due to having lower transpirational surface area, and having higher growth rates due to nutrient additions in readily available forms as urine and feces (Hobbs 1996, Table 3.2). The amount of compensation is maximized at intermediate rates of herbivory and declines as herbivory becomes very intense.

Compensation is important in maintaining stability in grazing systems. Some grasslands with migratory herbivores (e.g., the shortgrass plains of the Serengeti in Tanzania) have had high rates of herbivory for millions of years without showing signs of overgrazing, such as annual plant dominance or high proportions of bare ground. Stability will occur as long as the grazing does not occur too frequently, or is moderate in intensity if it is frequent, or if the herbivores migrate out of the area for part of the year to allow for recovery (McNaughton 1977). Grazing that is too frequent (chronic "overgrazing") without time for recovery (Oesterheld and McNaughton 1991) can override the ability of plants to compensate for tissue lost and can lead to declines in growth and fitness. Yet, grazing has clear negative effects in some systems, which suggest that sites vary in how they respond to grazing. In arid grasslands at the grassland-desert boundary, microbial crusts are broken by cattle hooves, annual grasses invade, and nutrient distribution becomes more concentrated under shrubs (Schlesinger et al. 1990).

The increases in nutrient content in foliage from grazing can attract repeated grazing on the nutritious regrowth of foliage. In some cases the grazing that results is extensive enough to create grazing lawns consisting of plants with low stature, dense placement of plant parts, and high allocation to leaves (McNaughton 1984). **Grazing lawns** are short-statured plant stands that are repeatedly grazed by herbivores. They are common in grazed grasslands. Short grass species such as *Cynodon dactylon* have adapted to this by having a short stature, which causes them to drop out of the community when grazing declines. The abundance of this species in the Serengeti is dependent on having intense grazing.

3.3.5 Effects of grazing on soil C

An important question is whether belowground processes such as root production are lower in areas where plants compensate for grazing by producing more aboveground biomass. If plants shift allocation from roots or crowns to shoots to recover from grazing, this could lead to less C entering soil C pools and eventually, to soil C loss. Crowns and roots have carbohydrate reserves, and their depletion could be involved in recovery. However, these shifts in allocation are not universal, and may depend on the evolutionary history of the species involved. In the Serengeti, which has a long history of intense grazing pressure, McNaughton et al. (1998) found no difference in root production inside and outside grazer exclosures. Wilsey et al. (1994,

1997) found that Serengeti species regrew aboveground by shifting allocation from stems to leaves after defoliation, whereas species from the flooding *Pampas* and Yellowstone National Park, where presumably grazing
pressure was lower historically than the Serengeti, shifted allocation from
roots and crowns to shoots. Isbell and Wilsey (2011) compared recovery
from cattle grazing between native North American prairie species and
non-native (mostly Eurasian) species. Non-native species recovered quickly
from grazing but produced less root biomass compared to plots that were
not grazed. The native species produced a similar amount of roots whether
grazed or not grazed. This could ultimately affect soil C accumulation over
time. Non-native species were able to recover from grazing more quickly
than native species, but at a cost to root production.

The effects of grazing on net ecosystem CO_2 uptake and ultimately, soil organic
carbon (i.e., C excluding inorganic carbonates) vary depending on the amount
of grazing and prairie type. Canopy level photosynthesis was higher and respiration was lower in grazed canopies compared to ungrazed canopies leading
to greater C uptake with grazing in one study (Wilsey et al. 2002), but the
photosynthesis and respiration changes cancelled each other out to result in
no differences with grazing in a second study (Risch and Frank 2005). Conant
et al. (2001) found in a large meta-analysis that moderate grazing led to a 2.9
percent increase in soil organic C on average, over no grazing and intensely
grazed treatments. Derner and Schuman (2007) found that short-grass prairie
had 24 percent higher organic carbon when grazed, but that organic carbon
was slightly lower under grazing in mixed and tallgrass prairie.

3.3.6 Herbivore response to plants

Herbivorous animals in turn respond to plants by showing selection for certain plant species and plant parts, by minimizing consumption of plants with
toxins, by evolving feeding strategies or larger body size, by evolving associations with bacterial symbionts, or by developing tolerance to plant defenses
in their host plants. Unpalatable plant tissues are often low in nutrients and
high in secondary compounds. Browsers have important adaptations to secondary compounds in dicots. An interesting example is found in white-tailed
deer, which feed often on tannin-rich acorns. Tannin binds proteins in the
diet, and the protein is passed through the gut without being utilized by the
animal. An intolerant herbivore could suffer from malnutrition by eating a
diet that is too high in tannin. Yet, acorns are a common food item in the
diets of deer, how do they do it? It turns out that they produce proteins
(e.g., proline) in their saliva (Hagerman and Robbins 1993). Proline binds
with the tannin in the saliva, causing it to pass harmlessly through the gut.

Important nutrients in plants for herbivores include N, and to a lesser extent Ca,
Na, and Mg. N in leaves can vary tremendously over time, with the highest N
concentrations occurring early in the growing season (Table 3.1). Older plant

tissues have lower N concentrations than young tissues. Thus, tissue that is regrowing following an earlier herbivory event has higher N concentrations than tissues before the event because the regrowing leaves are young. Leaf N is also higher, on average in species with the C_3 mode of photosynthesis than in C_4 species (Wilsey et al. 1997). However, this is "on average", i.e., is only found when leaves of comparable developmental state are compared. Because the green-up time is later in C_4 species (e.g., bluestems, indian grass, switchgrass), young tissues of C_4 species might have a higher N concentration than the much older tissues of C_3 species that are growing at the same time. Herbivores can optimize their intake of N by switching from C_3 to C_4 species shortly after the C_4 species have greened up. Many C_4 grasslands of the world are important for cattle (e.g., Konza Prairie) or wild mammal production (e.g., tropical grasslands).

Body size in animals is particularly important to their feeding ecology, especially when considering species within a feeding strategy (i.e., comparing within hind gut fermenters, or within ruminants for example). Body size is important when food quality is low. Because basal metabolism and body size are allometric (that is, metabolism scales with mass with a slope less than one, in this case approximately 0.75, Figure 3.7), it means that basal metabolism rises more slowly than body mass as size increases from small to large. The large size and large gut volume enables large body mass animals such as buffalo to maintain their metabolism on large amounts of low quality tissue. Small animals on the other hand, have a high metabolism per unit mass, have smaller guts, and as a result, have to select high quality forage. This can explain for instance, the herbivores preference for high quality green flushes that are found after fire (Figures 3.2, 3.3). Small herbivores show a higher preference for green flushes than large herbivores (Wilsey 1996, Sensenig et al. 2010). Large herbivores tend to select areas with higher quantity of food (unburned areas, and other tall statured areas), which tend to be of lower quality.

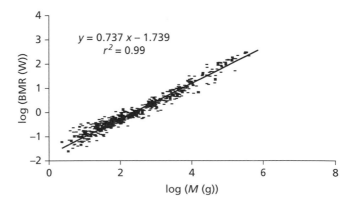

Figure 3.7   Basal metabolic rate of mammals (y-axis) plotted against body mass in grams. The slope is not significantly different from 0.75. (Used with permission from Savage et al. 2004 Functional Ecology).

3.4 Drought

Plant adaptations to drought include having high root–shoot ratios, short stature, and a conservative growth strategy (Grime 1977). Coughenour (1985) outlines the similarities in adaptations to herbivory and drought. Human selection for cultivars that are drought tolerant will probably lead to greater herbivory tolerance as well. Traits that evolved as adaptations to drought later became exapatations to herbivory, favoring plant recovery to both drought and herbivory (Coughenour 1985). Allocating growth towards roots and crowns (Figure 3.1) allows the perennial part of the plant to persist through the drought period. Being short-statured is usually associated with less surface area for water loss thereby lowering transpiration. Stress-tolerant plant species allocate a large proportion of their resources to storage over growth, have low growth rates in time periods when resources are plentiful, and have high allocation to structural and constitutive herbivore defense (Grime 1977, Coley et al. 1985, Chapin et al. 1990).

Drought can affect grasslands at a variety of time scales (Smith et al. 2009). At the individual plant level and the shortest time scale, the stress hormone absissic acid (ABA) can lead to stomatal closure during dry periods (Chapin 1991). Precipitation deficits and enhanced evaporative demand trigger soil moisture deficits, and this can lead plants to go into dormancy or die. There are many traits involved in drought stress, including an integrated stress response, conservative growth strategy, narrow leaves, and clonal growth (Chapin 1991, Craine et al. 2013). Among 426 grass species sampled by Craine et al. (2013), leaf width was an important predictor of drought response, with narrow-leaved species being more drought tolerant than wide-leaved species. Leaf width did not have a phylogenetic signal, and physiological drought tolerance was found throughout the grass phylogeny. The authors suggest that there are many grasses that are drought tolerant in grasslands throughout the world, and that maintaining grass diversity will help in drought recovery.

On a longer time frame, plants and animals must recover from drought with buds and seeds and egg hatching, respectively. Seedling establishment is surprisingly variable among years and locations in grassland species (Stuble et al. 2017). As a result, grassland species often rely largely on vegetative reproduction, producing new vegetative shoots (ramets) that are attached to the parent. Benson and Hartnett (2006) estimate that close to 99 percent of new grassland plant recruits are from vegetative reproduction, with only 1 percent of new recruits appearing as seedlings. Vegetative reproduction can allow rapid recolonization after droughts have ended.

At still longer time frames, drought can lead to species reordering, and eventually to species loss and immigration (Smith et al. 2009). These shifts and species loss can have long-term ramifications, especially if community shifts favor species with extreme traits. A period of dry years led to forb dominance in European grasslands, and this shift led to reduced net primary productivity for more than ten years following the end of the drought (Stampfi and Zeiter 2004). Weaver (1954, 1968) reported on grassland response to droughts during the Dust Bowl era in the USA Great Plains. He reported that many tallgrass prairies were dominated by the shortgrass species *Bouteloua gracilis* or *hirsuta* during the drought. The presence of these species stabilized biomass and prevented erosion during the drought. Several years after the drought ended, the tallgrass species increased in abundance to reclaim their place as the dominant species. This stresses the importance of rare species (or transient species of Grime 1998) in grassland response to drought. Rare species may not have important effects on ecosystem measures in average years, but can have important effects in extreme years (Isbell et al. 2011).

Drought can also result in reduced tree abundance. Trees can sometimes have high mortality during and after droughts, and the trees most likely to die during the drought are the ones that had low growth rates before the drought (Polley et al. 2016). Stress from droughts could therefore lead to lower tree cover in savannas and grasslands with woody plant encroachment.

Drought tolerance can also be important in biomass stability from year to year. Polley et al. (2013) found that the species that produced the most consistent biomass production between wet and dry years had a conservative growth strategy. In wet years, they produced a moderate amount of biomass compared to less stress tolerant species. In dry years, they had smaller reductions in growth compared to less conservative species. Wilsey et al. (2014) found that having dominant grasses that are stable stabilizes the community as a whole. Isbell et al. (2015) found that grassland experimental plots with the greatest numbers of species present were the most resistant to either exceptionally dry or exceptionally wet time periods.

3.5 Other extremes in precipitation

Relatively less work has been done on exceptionally wet periods compared to droughts. Some have suggested that wet years might be less important than droughts because net primary productivity saturates with high precipitation amounts. Precipitation above the peak will thus have little or no additional effects on NPP. However, in some cases, high precipitation periods could lead to shifts in species composition, for example from native species to exotic species, and these shifts could feed back to have large effects on species diversity and ecosystem functioning in the years following the shift to the new state. More research is needed on wet vs. dry periods, species shifts over long time frames, and drought legacy effects on grassland ecosystems.

4 Nutrient Cycling and Energy Flow in Grasslands

Grassland nutrient cycles and energy flows are well studied. This is partially because grasslands are so important for providing animal protein for human food consumption, and for storing organic matter and nutrients in soil for later crop production. A better understanding of energy flow through the herbivore and detritivore food webs can be used to enhance food production for humans. An understanding of how to manage grasslands for greater organic matter accrual will enhance our ability to take up and store CO_2 from our atmosphere.

4.1 Energy flow through grassland ecosystems

Energy flows from the sun, through primary producers, to consumers (primary, secondary, tertiary production), and much of the energy is lost as heat at each stage (Figure 4.1). Only ~1 percent of solar energy is converted to plant and algal tissue. Energy flow can be measured in units of energy such as Joules, but it is more often measured in units of mass or carbon. There has been much work on gross primary productivity (GPP), or the amount of solar energy that enters ecosystems and is converted to usable energy per unit time (weeks, months, or years usually). GPP is commonly considered by grassland ecologists because it can be measured from satellites as explained forthwith.

Net primary productivity (NPP) is the amount of C or biomass that accumulates over some time period. Photosynthesis by autotrophs leads to biomass accumulation, and autotrophic respiration leads to biomass loss. Photosynthesis uses solar energy to combine CO_2 and H_2O into carbohydrates. This material is then used in respiration, so:

NPP = C (or mass) gained from photosynthesis−C (or mass) lost from autotrophic respiration.

The Biology of Grasslands. Brian J. Wilsey. Published 2018 by Oxford University Press. © Brian J. Wilsey 2018.
DOI 10.1093/oso/9780198744511.001.0001

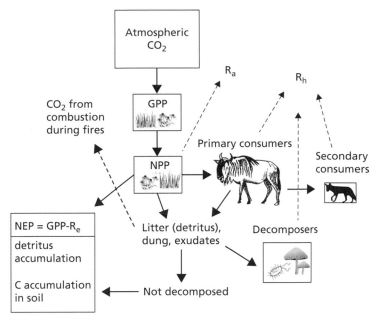

Figure 4.1   Energy flow in grassland ecosystems. CO_2 is fixed by the ecosystem as gross primary productivity (GPP), which is fueled by solar energy. Some of the carbon is respired by plants (autotrophic respiration Ra), and the remainder (GPP—R_a) is net primary productivity (NPP). NPP is consumed (H) or dies to produce detritus (litter aboveground, dead roots belowground). The portion that is consumed is converted into heterotrophic biomass, and after accounting for respiration is net secondary productivity and tertiary productivity. The portion of NPP that is not consumed or decomposed adds to NEP (net ecosystem productivity). Primary consumers, secondary consumers, and decomposers release CO_2 as heterotrophic respiration (R_h). NEP is measured as GPP—R_e, where ecosystem respiration $R_e = R_h + R_a$. Export and import of C from outside the grassland is not included in the figure and is assumed to be negligible in most grasslands. Areas that have migratory animals moving into and out of the site have significant amounts of exported and imported C, and this should be taken into account in these situations.

NPP is usually measured over a growing season time period, but it can be measured over shorter time periods as well. It can be measured by measuring carbon fluxes (Ruimy et al. 1995), but it is more commonly estimated with biomass accumulation over time with the equation:

$$NPP = (B_{t+1} - B_t) + D + H,$$

where B is biomass at time t, D is biomass lost through decomposition, and H is biomass lost to herbivores. Annual NPP is estimated by summing positive biomass increments across time periods during the growing season, adding in that lost to decomposers and herbivores. It is typically presented in units of g/m²/year or kg/ha/year. B is standing crop biomass, which is the amount of biomass present at any given time. Biomass is not the same thing as productivity. The confusing thing for some students is that biomass is used as a measure of productivity if it is measured as peak biomass and standard assumptions apply. Peak biomass is the amount of biomass that is found once a full canopy develops (max of Figure 4.2), but before significant decomposition has occurred. Peak biomass NPP is highly correlated with other estimates of NPP from approaches that sum biomass increments, but is typically lower overall due to some decomposition that can occur during the time period under study (Scurlock et al. 2002). Using peak biomass assumes that B_t is zero (usually in winter), and that D and H are negligible. Typically, peak biomass is used as a measure of NPP in grasslands where biomass hits zero in the winter or dry season and little biomass is lost from decomposition or herbivory during the study period. These are usually realistic assumptions in ungrazed grasslands, where biomass decomposition is typically low during the growing season. Peak biomass is perhaps the most commonly used measure of NPP in grassland ecology (Figure 4.2). In forests

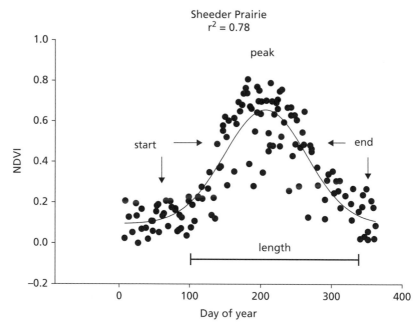

Figure 4.2   NDVI (Normalized Difference Vegetation Index) measured from space by the Landsat satellite for Sheeder prairie in Iowa, USA. Five years of data were combined from 30 x 30 m pixels, and cloudy days were removed before plotting. In this grassland, the data are unimodal (i.e., there is a single peak after a canopy develops). Green-up date, or start of season, is estimated as the day that NDVI rises above low winter values, or as the inflection point where NDVI is 50 percent of peak values (Wilsey et al. 2017). Senescence date, or end of season, is estimated as the day that NDVI drops to low winter values or as the inflection point where NDVI drops to 50 percent of peak values. These dates are sometimes expressed in growing degree days, which are based on accumulated temperature changes.

or deserts, the presence of evergreen shrubs and trees makes using peak biomass problematic.

In areas with grazing, offtake by animals (H) is also production, and this value has to be estimated and added back to estimate NPP. This makes estimating NPP a bit more involved in grazed situations. Most NPP estimates either assume that offtake to herbivores is minimal, or go to great pains to measure offtake with movable exclosures (McNaughton et al. 1996, Figure 4.3). Movable exclosure methods involve setting up exclosures for short time periods (usually a week to a month), clipping biomass inside and outside exclosures at the end of the time period after grazing has occurred, and then calculating offtake for that time period as biomass inside—biomass outside (McNaughton et al. 1996). Offtake (H) is then added to biomass produced to estimate NPP. This has been done in many grazed grasslands (McNaughton 1985, Frank and McNaughton 1992, Wilsey et al. 2002, Martin and Wilsey 2006, Frank et al. 2016).

Ignoring offtake by herbivores (H) can lead to erroneous conclusions about NPP. In some situations, such as comparing areas with moderate grazing to areas with no grazing, standing crop biomass can be lower but net primary productivity can be higher with grazing than without. Regrowing tissues after grazing can have higher N content with younger tissues that have higher rates of photosynthesis. Non-grazed areas can have reduced NPP due

Figure 4.3   A moveable exclosure used to measure off-take by grazing mammals. Biomass is clipped after a short time period (e.g., one week to one month) inside and outside the exclosure. Biomass differences between the two are estimates of consumption (H). Consumption is added to biomass accumulation to estimate net primary productivity (McNaughton et al. 1996). Photo: Brian Wilsey.

Figure 4.4   An herbivore exclosure in the Serengeti Plains, Tanzania shows the effects of grazing (outside) vs. no grazing (inside). Biomass and litter often accumulate like this in areas without grazing. In some grasslands, grazing mammals can consume greater than half of net primary productivity each year, and their activities can affect plant species composition, biodiversity and nutrient and carbon cycling. Photo: Brian Wilsey.

having older tissues and large litter buildups (Figure 4.4, McNaughton 1985, Knapp and Seastedt 1986, Wilsey and Martin 2015). In these kinds of situations, estimating offtake to grazers is especially imperative. Offtake to grazers (consumption) is often positively correlated with NPP (Frank and McNaughton 1992, Wilsey et al. 2002, Martin and Wilsey 2006). Positive offtake-NPP correlations can either be due to grazers seeking out more productive areas for grazing or due to grazing itself increasing NPP.

Typically, NPP is measured aboveground only, ignoring belowground production. This is denoted as aboveground primary productivity (ANPP). ANPP is not always the same as NPP in grasslands, where greater than half the production can occur belowground. Estimating belowground production uses ingrowth cores, which are cores of root-free soil placed into the field (Owensby et al. 1993). Since root biomass is zero at the beginning of the time period, the roots that have accumulated at the end of the time period are a measure of production. Other measures are the maximum–minimum method (McNaughton et al. 1996), which works if the minimum root biomass is near zero in the winter in temperate systems or during the dry season in tropical systems. The maximum is measured at peak biomass. Root minirhizatrons measure root elongation rate and width, and then this can be converted to production with regression equations. Belowground production is especially important to carbon storage in grasslands because roots have lower N content than leaves and stems, which leads to a greater proportion of roots resisting decomposition and being sequestered.

Biomass that is produced senesces and becomes litter (detritus) and the litter decomposes, losing mass. Decomposition occurs primarily in the shallow layers of soil of grasslands for aboveground litter and to a depth of 1 meter in the soil for roots (Figure 4.1). Detritus is processed by heterotrophic detritivores and microbes, respiring CO_2 and converting organic matter into humus or other recalcitrant forms of C (Figure 4.1). The amount of biomass or C that remains after decomposition is NEP, or net ecosystem productivity ($g/m^2/year$ or $kg/ha/year$). NEP can be estimated as:

$$NEP = GPP - C \text{ lost from ecosystem respiration (autotrophic + heterotrophic), or:}$$

$$NEP = NPP - C \text{ lost from heterotrophic respiration.}$$

Figure 4.5 Net ecosystem carbon exchange can be measured with Eddy Covariance towers or with chambers, pictured here in a grazed grassland. A shade cloth is on the chamber to measure net C uptake as a function of light availability (i.e., light response curves). Hoses are connected to a cooling system to keep temperatures inside the chamber at ambient levels (Wilsey et al. 2002). Photo: Brian Wilsey.

C loss from heterotrophic respiration is usually measured with decomposition, or the amount of NPP that is lost as mass loss over time. NEP can be estimated with carbon flux techniques (respiration is C loss) or litter mass loss. NEP and NPP (GPP – C lost from autotrophic respiration) are not the same thing, and the two terms should not be used interchangeably. NEP is important because if it is negative, it means that the site is a carbon source, or that loss from respiration is greater than that gained by NPP. If it is positive, the site is a carbon sink. NEP is commonly measured with chamber techniques, eddy covariance, and Bowen ratio towers (Ruimy et al. 1995, Figure 4.5). Understanding NEP requires measuring NPP as well as decomposition and understanding the factors that influence both.

4.2 Satellite measurements of GPP and NPP

Remote sensing, which is defined as sampling of the land surface without touching it, provides valuable tools for assessing GPP and NPP. Since about the 1970s, data on plant activity has been available to researchers from satellite measurements. Satellites are equipped with spectroradiometers that measure reflectance at multiple wavelengths. This has enabled us to develop a much more extensive, and even global scale view of what controls productivity. ANPP is commonly measured from space with the Enhanced Vegetation Index (EVI) or the **Normalized Difference Vegetation Index (NDVI)**:

$$NDVI=(NIR-VIS)/(NIR+VIS)$$

NIR and VIS are reflectance in the near infrared (730–900 nm) and visible spectral ranges (550–680 nm), respectively. ANPP can be estimated by measuring NDVI over time (Figure 4.2), and then integrating under the curve within the growing season. Xia et al. (2015) found that for most grasslands, multiplying the peak NDVI by the length of the growing season (ending date–starting date) provides an estimate of ANPP. This works best in sites that have a unimodal relationship with day of year (Figure 4.2). Bimodal distributions, which are more common in southern locations (Wilsey et al. 2017), are a bit more complicated because you have to fit data to two or more modes (peaks) in the data. The beginning and end of the growing season are found when NDVI begins to rise above low winter values (beginning, green-up), or when it drops back to low winter values (end, senescence, Figure 4.2). GPP, NPP, and NEP are key ecosystem measures because they represent the amount of energy entering the system, energy available to higher trophic levels, and the amount of carbon storage that occurs over time, respectively.

Remote sensing approaches do have limitations in grasslands. They are especially poor at estimating NPP in areas with grazing because they do not incorporate offtake to herbivores (H) in their calculations. Areas that are regrowing after grazing or haying have unusually high NDVI values that peak after each grazing or haying event, and the resulting curve is not uni-modal. Incorporating estimates of offtake by herbivores could be estimated to improve regional estimates of NPP.

Global estimates of NPP were used by Running (2012) to measure planetary boundaries of the earth. Planetary boundaries are variables that predict the planet's habitability (Running 2012), and include variables that address human caused changes including climate change, ocean acidification, land-use change, biodiversity loss, and alterations to the N and P cycles. Net primary productivity offers an integrative measure that might be an easily measured global surrogate for these measures (Running 2012). Global NPP is estimated to be ~53.6 Pg C and it varies by <2 percent annually. The pro-portion of NPP co-opted by humans was estimated in 1986 to be 40 percent of total (Vitousek et al. 1986), and this was updated to be 37 percent (20 of 53.6 Pg) by Running (2012). The amount of available NPP that is unavailable to humans might become a detriment to future human population increases (Running 2012).

4.3 Geographical variation in productivity

NPP varies regionally, and the important predictors of NPP are most often rainfall and temperature. Across grasslands on a geographic scale, NPP increases with annual precipitation (rainfall + snow and sleet) and average temperatures (Figure 4.6). Nitrogen mineralization (N available for plant growth that accumulates over time) also increases as precipitation rises, so the two increase in tandem (Burke et al. 1997). Soil organic carbon also increases with precipitation (Figure 4.7). Thus, NPP increases from semi-arid to subhumid grasslands due to greater water and N and C availability. Higher NPP with higher temperature is partially caused by a greater propor-tion of C_4 plant species in warmer areas, which are more productive than their C_3 counterparts (see Figure 4.8).

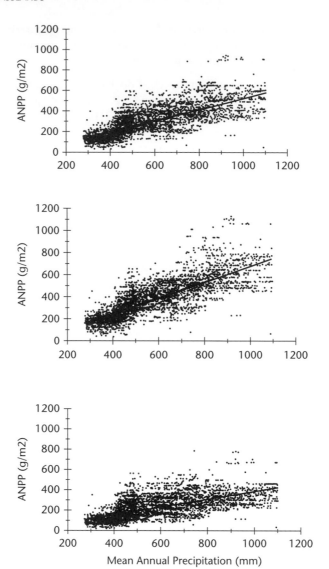

Figure 4.6   Relationship between peak biomass (a measure of ANPP) and precipitation amounts across grasslands of the Great Plains of North America (Lauenroth et al. 1999, used with permission) in normal (top, $r^2 = 0.56$, slope=0.53), favorable (middle, $r^2 = 0.66$, slope=0.71) and unfavorable (bottom, $r^2 = 0.43$, slope=0.38) conditions. N mineralization also increases with mean annual precipitation (Burke et al. 1997), so ANPP increases with joint increases in water and nitrogen availability. Organic carbon in soil (total C—carbonates) also increases with water availability (Figure 4.7, Derner and Schuman 2007).

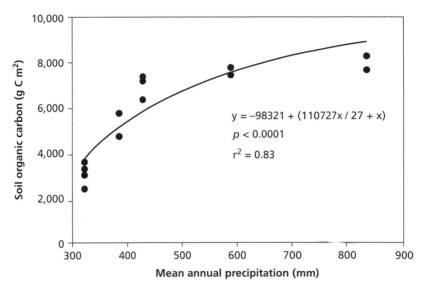

Figure 4.7   Soil organic carbon (total C—C in carbonates) increases with mean annual precipitation across grassland sites (Derner and Schuman 2007, used with permission).

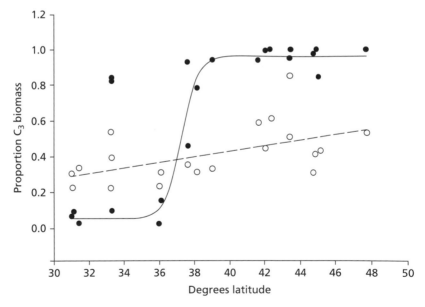

Figure 4.8   Proportion of aboveground biomass from plants with the C_3 mode of photosynthesis (1—C_4 biomass) in exotic- (closed circles, solid line) and native-dominated (open circles, dashed line) grasslands across a latitudinal gradient in the tallgrass prairie region, USA (Martin et al. 2014 Oecologia, used with permission). Sites span a gradient from northern Minnesota through Iowa, Missouri, Kansas, Oklahoma to central Texas. The proportion of C_3 biomass production increases as one moves north across native dominated sites. Exotic dominated sites are strongly dominated by either C_3 or C_4 species, which accentuates the gradient.

4.4 Within region productivity

Within a region, that is, among sites that have similar levels of precipitation and temperatures, NPP and NEP can be strongly affected by plant and animal species composition and diversity. Biodiversity-ecosystem function experiments on common soil types usually find orders of magnitude differences in NPP due to species compositional differences, and NPP increases with species diversity (see Chapter 6 for more details). The proportion of C_4 species in the plant community can be important as well. Grasslands with a high abundance of C_4 grasses in the south have higher NPP than areas with a lower abundance of C_4 grass in the north (Teeri and Stowe 1976, Epstein et al. 1997). The proportion of C_4 species partially underlies the temperature—NPP correlation.

The C_3–C_4 fraction of NPP can also be affected by the presence of non-native (exotic) species. Comparing 21 pairs of grassland sites at common latitudes across the tallgrass prairie region from northern Minnesota to central Texas of the USA, Martin et al. (2014) found that grasslands dominated by non-native species had more extreme proportions of either C_4 or C_3 species, depending on the species that were locally dominant (Figure 4.8). Native-dominated sites all had a mixture of C_3 and C_4 species, with a linear increase in C_3 species as one moved north. Sites dominated by non-native species were strongly dominated by C_4 species in the south, and strongly dominated by C_3 species in the north, with a switch point in Kansas. This resulted in NPP being higher in native than non-native dominated sites in the north, but higher in non-native dominated sites than native sites in the south (Martin et al. 2014). Increasing the proportion of C_4 species in a grassland experimentally can increase NPP (Foster et al. 2009).

4.5 Year-to-year variability in productivity

Net primary productivity also varies significantly from year to year in grasslands. Variation among years was higher in grasslands than it was in forests and deserts (Knapp and Smith 2001). This result was surprising to many when this paper was published, because many thought that deserts had the most year to year variability. Deserts responded less strongly than grasslands to wet years, which explained their smaller amounts of year to year variability. Large increases in NPP during wet years led to the highest CV in NPP in grasslands.

4.6 Water cycle

Globally, the water cycle is based on interactions between oceans and land via the atmosphere. Over the oceans, evaporation is greater than precipitation, and the excess water vapor moves over land, where precipitation is greater than evaporation (overall, with local exceptions explained subsequently). Water enters the grassland system as rain or sleet or snow (precipitation). It is lost through evaporation from soil surfaces, through transpiration through leaves, as percolation through the soil column to the water table, or as runoff across soil surfaces. Evaporation + transpiration is called evapotranspiration. If precipitation is greater than evapotranspiration, we call the grasslands humid grasslands. If evapotranspiration is greater than precipitation, we call them arid grasslands. On local scales, the amount of topography can be important to water movement, and in some dry, hilly grassland regions the locals joke that there are only two types of situations: drought and runoff.

Water strongly influences the N and P cycles, and water is limiting in many grasslands. Water plus N had greater effects on biomass than N alone in the studies of Harpole and Tilman (2007), DeMalach et al. (2017). The most common abiotic forms of N and P (nitrate [NO_3^-] and PO_4^{3-}) are highly mobile in water and can enter water-ways, leading to algal blooms and reduced water quality. Water links N and P in grasslands to grassland creeks and rivers, and eventually to marine systems.

The biota in turn can greatly modify the water cycle. Plants face an interesting dilemma: they have to take up CO_2 for photosynthesis through pores in their leaf surfaces (stomata). However, when they open their stomata, they lose water through evaporation from the leaf to the atmosphere (transpiration). Plants have adapted a remarkable array of ways to take up CO_2 without losing too much water. The amount of CO_2 fixed per unit of H_2O is termed water use efficiency (WUE). WUE varies across plant species and is higher in C_4 plant species than C_3 plant species. Crassulacean acid metabolism (CAM) species have high WUE because they open their stomata only at night. Plants can regulate their stomatal closure according to the concentrations of CO_2 and water vapor in the atmosphere; experimentally increasing CO_2 has been shown to lead to more closure of stomata, higher WUE, and increased moisture in the soil under plants (Polley et al. 2014). Litter mass, which as we have seen is modified by the amount of grazing, has been shown to greatly affect the amount of water that is lost to evaporation. Litter buildup in areas with low grazing intensities can affect light levels and soil temperatures (Knapp and Seastedt 1986), and ultimately, seedling recruitment (Wilsey and Polley 2003). Thus, the water cycle can be influenced by changes in the plant and animal community.

Liebig's Law of the Minimum suggests that there is a single limiting nutrient at any particular time. However, the great temporal variability that exists within a growing season can lead to multiple resource limitation. For example, early in the growing season, water, nutrients, and light can all be high, and growth is rapid due to having no limiting resources. After a canopy has developed, light can be the most limiting resource, at least to the species growing under the canopy. Grazing or human disturbances like haying can increase light availability, and limitation by other resources can become apparent. In many grasslands, N and P can be jointly limiting (Fay et al. 2015, Harpole et al. 2016), and additions of multiple nutrients leads to greater increases in growth and reductions in species richness than addition of a single nutrient like N. Water is almost always limiting in grasslands, but timing of water additions appears to be important.DeMalach et al. (2017) found in a meta-analysis that additions of water and N lead to greater biomass than additions of water or N alone. Surprisingly, addition of N caused a reduction in species richness, but additions of water did not. This was because water additions favored forbs over grasses, and N additions favored grasses over forbs. Grasses reduce species richness to a greater extent than forbs.

Moving up a trophic level, we can measure secondary productivity as the accumulation of herbivore or detritivore biomass over time.White (1983) found that secondary productivity was positively correlated with NPP across sites in Africa.McNaughton et al. (1989) found that herbivore consumption and secondary productivity were linearly related to NPP across the ecosystems measured. This suggests that NPP is a key regulator of (or is regulated by) animals and higher trophic levels (McNaughton et al. 1989). Interestingly, NPP is also positively correlated with herbivore consumption and productivity in livestock pasture systems of South America, but the intercept of the regression relationship is higher for South America than Africa (Oesterheld et al. 1992), suggesting that animal husbandry leads to systems with higher herbivore biomass and consumption rates.

4.7 N and P cycles

The nitrogen cycle is interesting because it has an atmospheric component and a major paradox. The paradox is that N is the most common gas in our atmosphere, making up 78 percent of the total, yet, it is commonly the most limiting nutrient to productivity. The explanation for this paradox is that N_2 in our atmosphere has a triple bond, which is difficult to break by almost all organisms in nature. N-fixation involves splitting this triple bond by *Rhizobium* sp. bacteria and a few species of cyanobacteria and converting N_2 to NH_4^+. Ammonium is then converted to NO_3^- (nitrification) in grasslands, which is the form of N most commonly taken up by plants. Rhizobium are found in nodules in roots of legumes and rarely, in some grass leaves. This ability to harbor *Rhizobium* makes legumes one of the most valuable genera in the plant kingdom to humans.

In wild grassland systems, N cycles internally, and the only external sources are from N fixation and lightning. Lightning can break the triple bond in N_2. Lightning adds only a trace amount of N to grassland ecosystems (Schlesinger 1997). NH_4^+ is adsorbed by soil particles at cation exchange sites in the soil, NO_3^- is not. NO_3^- is highly mobile and moves in solution through the soil column. Diffusion of these molecules is slow in soil. Roots grow through the soil taking up NO_3^- and NH_4^+, which are carried up to the rest of the plant in the transpiration stream where it is converted into enzymes or structural proteins. In arctic tundra, plants can take up organic forms of N (amino acids—NH_2). The N content of leaves is correlated with the rate of photosynthesis, which links (couples) the N and C cycles (Field and Mooney 1986). Leaves, stems, or roots are consumed by herbivores and herbivores are consumed by carnivores, moving N through the herbivore food web. When plants and animals die, N is passed through the detritivore pathway as litter (dead plant tissue) or carcasses of animals. The N content and C:N ratio of litter is important to detritivores, and decomposition rate is higher when the N content of the litter is high. Microbial decomposers break down organic forms of N (NH_2) into NH_4^+ and NO_3^- during mineralization. The conversion of NH_4^+ to NO^- is called nitrification. NO_3^- is important because of its highly mobile nature, which allows it to move through the soil column to ground water or across the soil surface into water bodies. NO_3^- can cause algal blooms and other water quality issues in water bodies. N returns to the atmosphere in reduced soils when bacteria use NO_3^- as an electron acceptor during anaerobic respiration or is volatilized as NH_3 gas after animal urination events during dry periods. Most N cycles internally without input by humans, but the natural cycle is now being over-run by human influences.

Since about the 1960s humans have been adding high amounts of N to agricultural fields in the form of synthetic fertilizer as NH_4NO_3 or as anhydrous NH_3. If it is added when plants are not active, it can be lost to the surroundings as NO_3^- or as the gas N_2O. The largest flux of N_2O to the atmosphere is usually found in late winter, early spring when soils thaw and warm up. N_2O is also released from the combustion process in fuel plants or cars. As a result, N is entering grassland ecosystems in rainfall at a rate of 0–10 kg/ha/year. It is also sometimes added to pasture systems as a fertilizer and is increased by planting of legumes. The result of all this human activity is the astonishing conclusion that humans are adding more N to the environment than all the natural forms of addition (fixation and lightning) combined (Vitousek et al. 1997).

The P cycle does not have an atmospheric component. In natural systems, P cycles internally. Weathering of rock provides the initial PO_4^{3-}, it is taken up by plant roots. P is a component of DNA, and RNA, and ATP, ADP, and ANP in plants and animals. Decomposers (fungi and bacteria) are important in the mineralization process, converting organic forms of P (litter and carcasses) back in PO_4^{3-} that can be taken up by plants. PO_4^{3-} is also being added to grasslands by humans in fertilizer and to water bodies as detergents, where it is causing water quality issues. P can be solely limiting to plant growth, especially in areas with high soil pH. P binds with Fe into unusable forms of P when pH is >7.5. P has also been found to colimit plant growth with N and water in some grasslands (Fay et al. 2015, Harpole et al. 2016).

Numerous studies have found that N and P additions are impacting grassland ecosystems. Studies are too numerous to list here. Most commonly, adding N experimentally leads to increased dominance usually by grasses, and a reduction in subordinate (rare) species (Suding et al. 2005). This was found to occur at N addition rates any higher than 14–25 kg/ha/yr in species rich calcerous grasslands of Europe (Bakker and Berendse 2002). Legumes are especially susceptible to N deposition, and they typically decline when grasslands are fertilized. Legumes also decline in tallgrass prairie regions that have encountered habitat loss and fragmentation (Leach and Givnish 1996). Non-native (exotic) species usually increase in abundance after N is added (Suding et al. 2005). As a result, N affected areas have reduced plant species diversity, a greater abundance of exotic species (Wedin and Tilman 1996) and altered species composition. This high N state is persistent, leading to a new state that may be hard to overcome even after fertilization is stopped (Isbell et al. 2013).

Animals, especially herbivores, can have strong effects on how plants respond to changes in resource availability. N additions reduced plant species richness by increasing plant growth and reducing light levels under canopies (Borer et al. 2014). Grazing mammals can counter this trend and can maintain high light levels even under N additions (Borer et al. 2014) thereby lessening negative effects of N addition in experimental plots from sites around the world. Moran (2014) found that areas that were grazed by bison had higher leaf-nutrient contents, and higher numbers of arthropods—specifically herbivores. Detritivores were not affected by bison grazing, as you might predict. N deposition can also cause outbreaks by insect herbivores, and these herbivore outbreaks can feed back to affect N addition effects (Moran et al. 2002). These results all point to the importance of herbivory in grasslands.

Over a longer-time frame, the increase in productivity from N additions can be lessened due to indirect effects associated with declines in species richness and diversity. Productivity can be positively related to species richness, and a reduction in species richness can lead to a reduction in productivity. So, over time, the reductions in species richness with N depositions can lead to a smaller productivity response than would be expected from N additions alone (Gross and Cardinale 2007). Further research is needed on how long-term changes in species diversity will feed-back to affect plant response to N additions, and where this mechanism is most prominent.

4.8 Fire effects on N and P cycles

Fires affect nutrient cycles. The N in litter is volatilized during a fire, and it does not accumulate in the ash like P does. However, after a fire occurs, the blackened soil is warmer due to its lower albedo; higher temperatures stimulate microbial activities, and this leads to short-term increases in N mineralization. Regrowing plant tissues are higher in N in burned than unburned vegetation. NPP is usually higher in the year following a fire. P does not volatilize, and phosphate accumulates in ash, which increases NPP in the year after fire in P-limited grasslands.

4.9 Importance of animals to nutrient cycles

In grasslands, herbivory by native species is more intense than in forests and deserts. Herbivores defoliate plants, removing and transporting nutrients and C with the animals to nearby areas, sometimes called "hotspots" (McNaughton et al. 1998). Removing plant biomass can lead to less canopy mass and lower rates of canopy respiration (Wilsey et al. 2002, Risch and Frank 2005). The remaining leaves can sometimes have higher photosynthetic rates, and they almost always have higher N concentrations. N concentration is positively correlated with photosynthesis rates in regrowing tissues. Higher N concentration can lead to repeated defoliation events when the herbivores return to defoliate the regrowing tissues. If this continues, a **grazing lawn** can form, which is dominated by short, high-N, fast-growing plant species that are tolerant to grazing (McNaughton 1984).

Grazing mammals can also have large effects on plants due to their tramping activities, their ability to transport seeds on their fur, and their conversion of low-quality N in plant tissue into high-quality N (more readily available to plants and microbes) in dung and urine. Bison and sometimes cattle and horses create wallows, areas where animals roll on their backs, which creates areas of soil disturbance in an otherwise homogenous grassland (Polley and Collins 1984). Fur of mammals is an excellent collector of seeds of all kinds, and seeds can be moved far from the plant on the backs of animals (Eyheralde 2015). Herbivores deposit nutrients, especially N in the form of urea when urinating, and they deposit C and N in dung.

Nutrient deposition by herbivores can lead to changes in nutrient cycling rates and increases in patchiness (and beta diversity in plants). N mineralization, a microbial process, has been shown to be increased by antelope and zebra grazing (McNaughton et al. 1997), and with prairie dog activities in colonies (Detling 1998). Urine hits cover one third of the short-grass plains of the Serengeti every year, which means that each patch of ground will receive urine about once every three years (McNaughton 1985). Plant N content is higher in the growing season after a urine hit, and grazing mammals are more likely to graze on this nutritious regrowth than in non-hit areas (Jaramillo and Detling 1992). These patches can have different plant species composition compared to other areas (Steinauer and Collins 1995).

Finally, when animals die, they can create high-quality food for scavengers and decomposing microbes, leading to nutrient-rich and disturbed patches that contribute to patchiness in grasslands (Figure 4.7, Figure 4.9).Towne (2000) followed bison, cattle, and deer carcasses in grasslands at Konza Prairie in Kansas, USA for five years. He found that inorganic nitrogen concentrations were significantly elevated the year after death. Plant communities become more dominated by forbs, and C_4 grasses declined where the carcass was located. The effects on the plant community lasted for five growing seasons, and perhaps more. These carcasses led to community heterogeneity and long-term spatial complexity.

Invertebrate species are important to ecosystem processes as well. Many invertebrate animals provide a link between fresh plant materials and bacteria by breaking down large plant parts into smaller particles that can be utilized by bacteria and fungi. Shredding species like Colembola, termites in

Figure 4.9   Marabou storks are major scavengers in African grasslands. They are among the largest flying birds, with wing spans of 225–287 cm. Photo: Brian Wilsey.

Figure 4.10   Termites are major litter shredders in tropical regions. Their mounds create spatial variability. Photo: Brian Wilsey.

the tropics, and some earthworm species shred plant materials into smaller fragments (Figure 4.10). Decomposition rates are increased when the diversity and evenness of shredding species is high vs. low. Bacteria and fungi mineralize organic forms of nutrients from plant roots and shoots, and animal carcasses into inorganic forms of nutrients that can then be taken up by plants. Most members of the Eubacteria clade are undescribed to science, and only a small percentage of grassland bacteria have been cultured. However, this will change over time as metagenomics tools are developed and refined to identify key genes in the soil. Perhaps most interestingly, dung beetles are important in processing dung in grazed grassland areas, thereby speeding nutrient cycling (Nichols et al. 2008). Plant production declines when dung beetles are removed experimentally.

As a result of the effects of plant species, plant functional groups and animals, the important regulators of grassland C and nutrient cycling shift from precipitation, temperature, and soil type across grasslands, to biota and slope within grasslands.

5 Trophic Cascades in Grasslands

5.1 Introduction

The previous chapter on energy flow (Chapter 4) presented a somewhat bottom-up approach to ecosystem structure. Bottom-up effects can be altered by top-down processes, and the two can interact in interesting ways (Borer et al. 2014). **Trophic cascades** occur when top predators have top-down effects on the next lower trophic level, which in turn affects the next trophic level below (i.e., effects cascade down). By reducing their prey abundances, predators can have an indirect effect on their prey's prey. Because of this cascading effect, it matters how many trophic levels are present in the system under study (Fretwell 1977). For example, primary producers will increase from predation in a three-trophic level food chain during a cascade event—adding a top predator at the third trophic level will reduce prey at the second level, which will lead to increased abundance of primary producers at the first level. In a four-level food chain, the primary producer level will decrease by predation during a cascade event. The same is true in a two-level food chain under trophic cascade theory. Trophic cascades are well known from aquatic systems, but they have also been studied in grassland systems. In this chapter, I will present three examples of top-down effects in grasslands: reintroduced wolves (*Canis lupus*), spider effects in human-derived grasslands (old-fields), and the effects of multiple predators in the greater Serengeti ecosystem.

Trophic cascades are examples of **indirect interactions**, which are important in ecology.

A predator-prey interaction is a direct interaction where the prey species is consumed by the predator. If the predator is a top predator in a three-trophic level system, the predator can have an indirect effect on the lowest trophic level by affecting the second trophic level. Predators can be classified into multiple functional groups, with two being commonly used: sit and wait predators and active predators. **Sit and wait predators** wait for prey and

The Biology of Grasslands. Brian J. Wilsey. Published 2018 by Oxford University Press. © Brian J. Wilsey 2018.
DOI 10.1093/oso/9780198744511.001.0001

ambush them when they arrive. **Active predators** (sometimes called stalkers) actively pursue their prey. Predators can also be generalists, eating a wide variety of prey, or specialists, often specializing on prey of a certain size. I will now discuss several examples of indirect interactions that can be important in grasslands.

5.2 Wolves in the northern range of Yellowstone

Wolves are top predators in temperate grasslands, and they can have important direct and indirect effects on grassland ecosystems. Coyotes and red foxes are smaller dog species (mesopredators) that interact with wolves. Typically, coyotes are more common in areas without wolves, and red foxes are more common in areas without coyotes (Levi and Wilmers 2012). Wolves evolved in Eurasia, and spread to North America about 750,000 years ago, establishing populations in a wide variety of habitats, including grasslands. Wolves were extirpated by humans in many parts of the world, but they are being reintroduced or are recolonizing protected areas in the USA and Canada. There are discussions of whether to reintroduce wolves into the UK and other parts of Europe. Wolves are sometimes seen as competitors with humans for wildlife species, and wolves prey on livestock, so reintroduction is often controversial.

Predators have been shown to have interesting effects on herbivore behavior, and they can alter species composition of prey, which has important top-down effects. In the northern range of Yellowstone National Park (Figure 5.1), which is a montane grassland, wolves were reintroduced in 1995. Concurrent with their introduction, the park made a concerted effort to encourage all predator populations, and mountain lions and grizzly bears have increased in concert with wolves. These three predators have had major effects on the prey animals in the park (Beshta and Ripple 2016). Prior to reintroduction, elk (*Cervus canadensis*) were by far the most abundant ungulate species, with bison (*Bison bison*) being less common. Aspen and willow recruitment in riparian areas was low, which was negatively affecting beaver populations. Wolves preferentially feed on elk over the much larger bison. Twenty years later (2015), elk were much less abundant, and bison were more common than elk (Frank et al. 2016). Bison grazing was heavier than the previous elk grazing, and the compensatory response of plants was smaller than it was in the past (Frank et al. 2016). Thus, it appears that wolves have altered the species composition of the grazing mammal community, with important implications for plant recovery from grazing.

Elk behavior has changed due to the reintroduction of wolves as well, and elk now feed more often away from wooded areas around streams. As a result, they are browsing less on woody plants (remember that elk are mixed

Figure 5.1   Northern range montane grassland in Yellowstone National Park, USA. Elk and bison are major herbivores in this grassland part of the park. Photo: Brian Wilsey.

feeders, Figure 5.2). The change in habitat use from a preferred site to a less preferred site due to the presence of predators was termed the "**Ecology of Fear**" by Brown et al. (1999) and Ripple and Beschta (2004). The presence of predators can have negative effects on animal reproductive rates, even without the predator consuming the prey (Preisser et al. 2005). Preisser et al. (2005) reported on studies with invertebrates where researchers mimicked the effects of predators by sewing the predators' mouth shut, and then placing them with prey. Prey species with these incapacitated predators had lower growth and reproductive rates than prey without them. This ecology of fear is an indirect effect of predators on prey and is probably widespread in nature.

The lower browsing pressure on tree seedlings in riparian areas due to elk predator avoidance behavior is expected to have long-term implications to riparian systems and the animals in those areas, including beaver (Beshta and Ripple 2016, Marshall et al. 2013). If trees can achieve 2 m height, they can grow above the browse line and survive the winter. Beaver rely on aspen and willow for forage and dam building, and ponds in dammed creeks support many other animal species. Managers in Yellowstone introduced beavers to parts of the northern range shortly after the wolves were introduced, making it difficult to determine if wolves are affecting beaver through an indirect interaction (Marshall et al. 2013). Marshall et al. (2013) did a rigorous experiment that compared human constructed dams that altered hydrology to controls, browsing by elk to controls (exclosures), or both treatments to controls. They found that both beaver ponds and browsing, but not reduced browsing alone, enabled willows to reach the height of 2 m. Thus, both

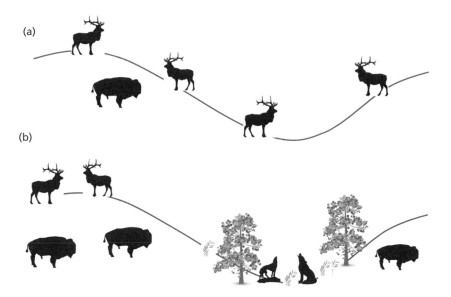

Figure 5.2   Ecology of fear at Yellowstone National Park (USA). Before wolves were re-introduced in 1995, elk grazed the uplands and browsed woody plants in riparian areas (A). After wolves established, their habitat use shifted towards uplands. Responses included an increase in bison, and growth of aspen and willows in riparian areas (Frank et al. 2016). Joint reintroduction of beaver to riparian areas has helped woody plant recruitment (denoted in an exaggerated way by the tall trees) by increasing the water table height.

beaver and a reduction of elk browsing were critical to recovery of riparian plants.

Interesting interactions occur among predators as well. Grizzly bears are much larger than wolves and steal food from wolves in Yellowstone. The much larger lions steal food from cheetahs in Africa in a similar manner. Some have hypothesized that lions have caused declines in cheetah populations. This sets up a predator hierarchy based on size that results in interesting indirect interactions.

5.3 Invertebrate predator effects on old fields

In old fields of eastern USA (human derived grasslands), invertebrate predators are filling a similar ecological role as wolves. Adding predatory praying mantises to old fields led to reduced herbivore loads on plants and increased plant biomass compared to control plots (Moran et al. 1996, Figure 5.3).

Sometimes "sit and wait" predators can change herbivore feeding and cause them to eat a lot of non-preferred foods if their preferred foods are in areas that are vulnerable to predation. Schmitz and colleagues have done a series

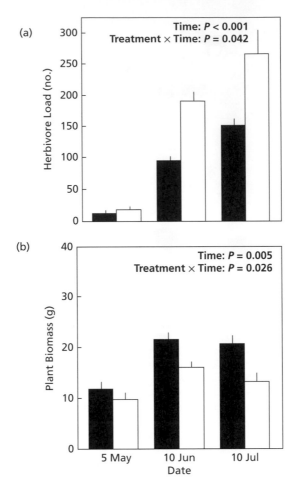

Figure 5.3   Results from an experiment that experimentally added predatory praying mantises to grassland plots (Moran et al. 1996). Solid bars are plots treated by adding mantises, and open bars are control plots. Top panel is the number of herbivorous insects, and the bottom panel is plant biomass per plot. Adding mantises reduced the herbivorous insect load and increased plant biomass, and this effect was larger as the growing season progressed.

of spider removal studies, and have found that spiders can increase plant-species diversity, especially the evenness component, by reducing plant dominance. The fields are strongly dominated by goldenrod species *Solidago canadensis* and have many grasshoppers without spiders (Figure 5.4). The herbivorous grasshoppers preferentially feed on grasses, reducing their abundance, which leads to stronger dominance by goldenrods in areas without spiders (Schmitz 2003). Adding spiders experimentally back into plots caused grasshoppers to move from grasses to escape predators, feeding on goldenrods instead of the preferred grasses. Feeding on goldenrods reduced their dominance, leading to increased species diversity of the plant community

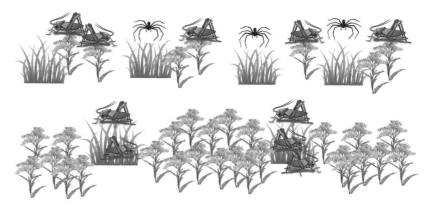

Figure 5.4   Spiders with the sit- and wait-strategy of hunting have top-down effects on plant diversity, specifically evenness, by having an indirect effect on the dominant species goldenrod (*Solidago canadensis*) in New England grasslands (USA). In the top panel, spiders on grasses caused grasshoppers to shift their feeding to goldenrods, which reduced their dominance significantly (Schmitz 2003). Because goldenrod is highly dominant in this system, grasshopper feeding leads to an increase in plant diversity as other species are released from competition.

overall. Later results suggested that this was due to the spiders being sit and wait predators. Ambush predatory spiders had different effects on grasshoppers.

These top-down effects by spiders can be affected by large mammal grazing. Zhong et al. (2017) found that adding sheep to a spider-grasshopper-grassland system shifted regulation from top-down to bottom-up in Chinese steppe meadows. Sheep are considered ecosystem engineers (Derner et al. 2009) that consume forbs in this system. Without sheep, forbs were abundant, which kept the dominant grass (*Leymus chinensis*) at lower relative abundances, which enabled grasshoppers to be regulated by spiders. Sheep preferentially fed on forbs, which favored the dominant grass *Leymus chinensis*, which had the effect of lessening the top-down regulation by spiders. Including cattle in the equation might complicate matters further if cattle reduced the biomass of *Leymus chinensis*. Further research is needed in how top-down vs. bottom-up regulation is affected by multiple herbivores in a realistic setting (Borer et al. 2014).

5.4 Serengeti, Ngorongoro

Perhaps the most well-studied predator-prey grassland system is the Greater Serengeti ecosystem of Tanzania, Africa. The Serengeti is located along a huge precipitation gradient running from northwest (wet) to southeast (dry) (Sinclair and Arcese 1995). A soil fertility gradient runs in the opposite direction. It is super-diverse in terms of ungulate mammals, with 28 species coexisting in the park. The 28 species provide a large sample size for testing hypotheses about body size and predation. The most common migratory species is the wildebeest (Gnu), which was described as a keystone species by Sinclair and Arcese (1995). The Serengeti Plains is a large treeless area in the southeastern part of the Serengeti National Park. Over one million grazing mammals concentrate their grazing in the plains region during the wet season, with wildebeest being most common. There are also large populations of zebra and gazelles (Grant's and Thomson's), as well as many other species (Figure 3.2). The apex predator community is largely intact, with large populations of lions, hyenas, and cheetahs (Figure 5.5). Wild dogs (*Lycaon pictus*) are recovering from formerly low numbers. Sinclair, Mduma and colleagues (1999) have studied whether ungulate populations are limited by food (bottom-up), or by predation (top-down). There are two important aspects to this system that differ from the Yellowstone system described previously: 1) the herbivores are migratory and move through the entire Serengeti each year, and 2) lions, the main predators are territorial. Thus, a lion territory can be surrounded by many thousands of prey animals during some parts of the year, and few or none during other parts of the year. Nearby, the Ngorongoro crater, which is a large volcanic crater (but much smaller than the Serengeti Plains) with the roughly the same set of animal species, was used by Sinclair and Mduma as a comparison. The animals are not migratory in the crater, and the entire area has lions and hyenas.

Figure 5.5 Spotted hyenas are major predators in the shortgrass Plains of the Serengeti National Park, Tanzania. The individual in the foreground has recently fed, and has a distended belly. Spotted hyenas travel long distances to find animal herds and return after feeding to their pups in dens. Photo: Brian Wilsey.

Surprisingly, Sinclair and colleagues found that the ungulates in the Serengeti plains are limited by food using a 40-year record (Mduma et al. 1999). Common causes of mortality were malnutrition, falling, and animals getting separated from the herd. Errant individuals are preyed upon, but the large herds appear to be more limited by food than by predation. The migration is associated with spatial variation in limiting nutrients P, Na, and N, and animals move to maximize intake of these limiting nutrients (McNaughton 1988). In the Ngorongoro Crater, and in other areas with resident (non-migratory) ungulates, the predators can regulate prey species (Sinclair and Arcese 1995). These results suggest that predators are more likely to limit their prey populations in non-migratory than migratory ungulate systems.

Although Serengeti plains ungulate populations appear to be regulated by bottom-up processes and not by predation, they do show adaptations to predation. Sinclair et al. (2000) suggests that the migrations themselves are adaptations to predation, allowing the ungulates to limit the effects of predators by moving seasonally. Migratory behavior may increase evolutionary fitness in prey animals, and the greater abundances of migratory populations than non-migratory populations in the Serengeti supports this view.

Interestingly, the phenology and synchrony of births vary among species in a way that is consistent with an adaptation to predation. Food quality is determined by crude protein, which can peak in the early part of the rainy season (Table 3.1) and food quantity is determined by the amount of biomass (food) that is available for consumption, which peaks towards the end of the growing season. Small ungulates tend to give birth before the peak in protein, so that food *quality* will peak shortly after birth, but when food quantity will be low. Large ungulate species give birth after the peak in protein, when food *quantity* is high, but food quality is low.

Synchrony of births, or having all calves drop close to the same day is a function of how precocious the young are, with the precocious wildebeest, topi, and buffalo having highly synchronous birth dates to satiate the predators in the area. **Predator satiation** is when prey species produce a large number of offspring all at once to overwhelm predators, which ensures that a greater number of prey will survive. After a few days, the precocious calves of these species can keep up with the herd. Non-precocious species ("hiders") such as impala, Thomson's and Grant's gazelles, which are unable to run from predators after birth, have asynchronous births (Figure 5.6). These non-precocious species spread out their birth dates in order to make their offspring less apparent to predators (Sinclair et al. 2000).

Top-down forces, and the ecology of fear are also important to tree species composition in African savannas. Impala browsing of *Acacia* species led to an increase in a species (*A. etbaica*) with especially long thorns, and a decrease in a thorn-free species *A. brevispica* (Ford et al. 2014). Impala prefer thorn-free plants, which was verified by experimentally adding and removing

Figure 5.6 Baby gazelle hiding in open grassland. Gazelles have non-precocious young, and they have staggered births as an anti-predator adaptation. In contrast, grazing mammals that have precocious young have synchronized births to satiate predators. Photo: Brian Wilsey.

thorns from both species. Impala behavior was impacted by predators, and browsing was higher in open glade areas that had fewer predators (leopards and wild dogs) than in more wooded areas. Several years after impala were removed with exclosures, the thornless species *A. brevispica* increased in abundance compared to the thorned species (Ford et al. 2014).

So, based on these examples of top-down control, are grasslands limited by top-down or bottom-up forces? Adding limiting nutrients such as fertilizer clearly has large effects on lower trophic levels that in turn, cascade up the food chain (see Chapter 4). How can bottom-up and top-down effects both be important in grasslands? Moran and Scheidler (2002) did a fertilizer and predator (spider) removal experiment simultaneously. Both had significant effects. This supported the hypothesis that bottom-up and top-down effects are not mutually exclusive, but that they interact. They summarized their results as: bottom up effects form the *potential* net primary productivity of the site, and top-down effects form the *realized* production of the site. Both effects are operating. After many decades of research on whether grasslands are limited by top-down or bottom-up forces, most ecologists have concluded that they interact.

6 Biodiversity and Ecosystem Functioning in Grasslands

6.1 Introduction

Native grasslands can be surprisingly diverse in species. For example, the early pioneers in the USA marveled at the huge diversity of wildflowers in the grasslands of the Great Plains. A wide variety of colors, heights, and patchiness greeted the first settlers, and they often commented on the prairie's beauty in their journals. Native grasslands contained a great number of grass and forb (wildflower) species, and the patchy nature of ungulate grazing and fires led to a complex landscape (Figure 1.1). This complexity was reduced after settlement and the decades of human activities that followed (Figure 1.1). After settlement, grasslands were managed as hayfields or pastures to produce forage for animals, and they were viewed from more of a utilitarian framework. Non-native species were introduced, and management tended to be applied in a uniform way across the landscape (Fuhlendorf and Engle 2001), which reduced species diversity at most scales. Synthetic fertilizer applications and atmospheric N deposition caused additional species losses (Silvertown et al. 2006, Suding et al. 2005). Beginning in about the mid-twentieth century, ecologists began to ask whether reductions in species diversity such as these have led to a greater frequency of invasions by insect pests and pathogens, more variable forage (biomass) production, and lowered resistance to droughts. Early work indicated that diversity and stability are positively related, which would mean that there may be management trade-offs involved when diversity is reduced due to managing for a single dominant species. By the 1990s (Schulze and Mooney 1994, Naeem et al. 1994), evidence began to accumulate that primary productivity could also be lower with reduced biodiversity. The controversy that followed these initial studies (e.g., Huston 1997) led to hundreds of studies with many forms of manipulation in a variety of ecosystems. Most of the early studies on this

The Biology of Grasslands. Brian J. Wilsey. Published 2018 by Oxford University Press. © Brian J. Wilsey 2018.
DOI 10.1093/oso/9780198744511.001.0001

topic, and many of the most influential since then have been conducted in grassland systems.

But first, what is biodiversity, and how do we go about measuring it? Biodiversity is a measure of variety of life forms, and can be assessed at the genetic, species, and landscape levels. Diversity can be higher with moderate levels of grazing when the grazer preferentially feeds on dominant species (Olff and Ritchie 1998), partially because grazing increases equalizing mechanisms of coexistence (Chesson 2000). When herbivores feed on rare species, such as the case of white-tailed deer, they can reduce diversity (Waller and Alverson 1997, Còté et al. 2004). In grasslands, rare plant species tend to be forbs (satellite species) and common species tend to be graminoids (core species, Collins and Glenn 1990).

6.2 Diversity measures

Species diversity is commonly measured in ecological studies. This is sometimes called community diversity, but we are usually talking about the number and relative abundances of species and not the number of communities. The number of communities is beta diversity. Species diversity can be partitioned into its basic components of richness (number of species) and evenness, and into spatial components as we will discuss forthwith. Functional diversity can be quantified by replacing species with functional groups, or trait values can be used in a variety of functional diversity measures. Species richness is the number of species in an area and is sometime called species density in cases where the number of species per unit area is being considered. Richness is a fundamental measure of communities, because it is a measure of the number of species that can potentially interact. All species can potentially increase from rare if they are present in the community. If they are not present (i.e., locally extinct), they must reinvade from elsewhere, which can take a long time. However, the presence-absence nature of richness can sometimes miss differences in diversity that exist across sites because it does not take into account species rarity and dominance. Evenness is a measure of how relative abundances vary across species, with a high-evenness site having roughly equal abundances and a low-evenness site having strong dominance by one or a few species. For example, an area with ten species with equal abundances is going to appear more diverse to us than an area with ten species with strong dominance by one species (90 percent of the total), and nine species with 1.1 percent. Rarity or high dominance can precede local extinction or immigration events. Abundances can be measured with the number of individuals, biomass or cover, or frequency (Collins and Glenn 1990). Local extinction rates are often higher in situations where evenness is low due to low abundances in rare species (Chapin et al. 2000, Wilsey and Potvin 2000, Smith and Knapp 2003, Wilsey and Polley 2004, Hillebrand et al. 2008). Evenness can also impact richness because the amount of rarity in a system will increase the probability of not detecting a species that is present (Brose et al. 2003). (Note: evenness is also called 'equitability'.) Because it measures abundances and not presence or absence, evenness can be more strongly correlated with the amount of rarity than species richness (Wilsey et al. 2005). Diversity measures that take into account evenness are important to consider when most species are rare and only a few species are abundant (Gaston 2010), i.e., in situations where dominance is especially high. It is less important to consider when species are close to being equally abundant, and richness is regulating other diversity measures (Wilsey and Stirling 2007). Surprisingly, measures of species richness and evenness are not always positively correlated (Stirling and Wilsey 2001, Wilsey et al. 2005, Ma 2005), and can be independent from one another, or even negatively related in some cases. Other metrics such as the Shannon and Simpson's diversity indices are compound measures that take into account both richness and evenness (Jost 2007). Phylogenetic diversity measures take into account the length of time that different clades (species usually) have been separate from one another (Cadotte et al. 2012). Cadotte et al. (2012) found that plots planted with high phylogenetic diversity have higher productivity than plots with low phylogenetic diversity, and they suggested that this measure can effectively get at the importance of functional diversity to ecosystem processes.

Which diversity measure to use depends on the hypotheses being tested. Usually richness is used when immigration and emigration and meta-community dynamics are being studied, animal communities are the focus, and longer timeframes are being considered (Wilsey and Stirling 2007). Compound measures and associated evenness measures are more often used when relative abundance are unequal (dominance exists), for plant communities, and when more detailed measurements are necessary (Stirling and Wilsey 2001, Wilsey et al. 2005). Using diversity components instead of single measures can sometimes lead to deeper insights on how diversity is being affected by the environment (Ma 2005).

Species diversity and richness measures can also be partitioned into spatial or temporal components. This enables us to study how diversity is altered as scale changes. Spatial components are called alpha if they are *within* patch or site measures, and beta when they are *across* patch or site measures. A measure of gamma diversity can be calculated by multiplying beta by alpha (Whitaker 1960). In recent years, beta diversity has been defined as a general measure of species turnover from location to location, and alpha as a mean within-location measure. Beta can be studied within a field from quadrant to quadrant or can be studied across fields or sites. Thus, beta can be estimated at multiple scales (beta$_1$, beta$_2$, etc., Lande 1996, Crist et al. 2003). Beta diversity is higher in situations where assembly history is different among patches (Martin and Wilsey 2012, Martin and Wilsey 2015), where grazing mammals are moderately grazing an area (Steinauer and Collins 1995), where there is great variation in soil types and topography (Harrison et al. 1992), and where dispersal rates are low. High dispersal between sites will reduce beta diversity (Loreau 2000).

A complementary approach to studying how scale influences diversity is to use species-area curves, or the number of species plotted against area (Williams 1964, McArthur and Wilson 1967, Rosenzweig 1995). Beta diversity is sometimes quantified as the slope of species-area curves. A rich literature exists on species area curves, and this approach is an effective way of scaling up species richness to larger areas. Williams (1964) and Rosenzweig (1995) discussed the different processes affecting the species-area slopes as one moves from local, to regional, to global scales. Plant species area curves do not level off, but continue to rise at various rates until the level of the globe is reached (Williams 1964).

6.3 Relationships between biodiversity and ecosystem functioning

The influential book on biodiversity and ecosystem function edited by Schulze and Mooney (1994), and many of the studies that were published shortly thereafter (e.g., Lawton 1994, Naeem et al. 1994), offered hypotheses on how biodiversity might be related to ecosystem functioning (Figure 6.1). The rivet hypothesis (Ehrlich and Ehrlich 1981) suggested that the relationship would be linear, where each species lost would have a corresponding reduction in ecosystem functioning (Figure 6.1). Diametrically opposed to this was the idiosyncratic hypothesis, which suggested that there would be no relationship between species richness and ecosystem functioning. Rather, it would vary as a function of the dominant species present (Lawton 1994, Naeem et al. 1994). The redundancy hypothesis predicted that ecosystem functioning would start to saturate after one species was present (Figure 6.1). Additional species beyond one would have no effect on ecosystem functioning. An intermediate view was that ecosystem functioning would increase with additional species, but the effect would be a saturating function (Vitousek and Hooper 1993). This framework was popular during the early years, and results supported one or more of these hypotheses depending on the response variable being considered. The early and highly influential study by Naeem et al. (1994) found that plant productivity was a saturating function of planted species' richness, but

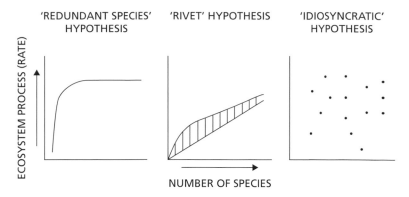

Figure 6.1   Three hypothetical relationships between ecosystem process rates and species richness (modified from Lawton 1994, used with permission). The redundant species hypothesis (left panel) predicts that ecosystem process rates will saturate very quickly as the number of species increases. Species above a small number are redundant in their effects. The rivet hypothesis (middle) predicts reduced function as the number of species declines. The idiosyncratic hypothesis predicts no relationship (right).

that many biogeochemical cycle measures had an idiosyncratic relationship with richness.

Many experimental and observational studies have been done on how ecosystem process rates will be impacted by reductions in biodiversity in the years since Schulze and Mooney's book came out in 1993. Although many ecosystem processes have been considered (Hooper et al. 2005), primary productivity has been, by far, the most often measured variable of ecosystem function. (For further details on other ecosystem processes, see Chapter 4). Meta-analyses most often analyze primary productivity relationships for this reason. Cardinale et al. (2007) found with a meta-analysis of many experimental studies that there is a general saturating relationship between diversity and net primary productivity. NPP increases experimental plots with higher diversity, usually with a peak being reached at some intermediate level. Because of the rich data set on this simple relationship, there has been a lot of interest in uncovering what lies behind it. It is not controversial among grassland researchers that this relationship is commonly observed, but there is some disagreement about the mechanisms behind the relationship. Understanding the mechanism is key to applying the research findings to real field situations.

The mechanism behind the positive relationship between diversity and ecosystem process rates can be due to at least four processes: 1) the species sampling effect, 2) the selection effect, 3) complementary resource use, or 4) pest outbreaks in low diversity plots. The sampling effect is the increased probability of including a species with extreme traits in species-rich plots. This is most relevant in studies that use random draws to establish richness levels. When study plots are set-up with different numbers of species, the species are usually randomly drawn from a larger species pool by investigators. Assuming that there is variation among species in their growth rates in the larger species pool, the process of drawing a greater number of species for species-rich plots will result in a higher probability that species with high productivity will be drawn. With a greater number of species selected, it is more likely that a species with extremely high growth rates will be drawn compared to plots with fewer species. This effect can lead to higher productivity even without any other process being involved. This is primarily an issue with studies that use random draws of species to establish species richness treatments. Random draw experiments have their strengths and weaknesses. Establishing non-random draw experiments can sometimes limit their generality and random draws are intended to enhance generality. However, most ecologists agree that most extinction in nature is not random, and that the results have to be used carefully to infer patterns in nature (Srivastava and Vellend 2005). Fortunately, we have now had many studies on both random and non-random extinctions to address this issue.

The second and third mechanisms, the selection and complementarity effect, are related to the concept of over and underyielding in mixture. Interactions can change between single species monocultures and more diverse multi-species mixtures. De Wit and coauthors (1966) suggested that it is informative to compare how species yield in mixture compares to what would be expected from single species monocultures. (Yield is usually biomass produced, but it can also be seed production or related measures.) As an example, with two species mixtures of equal abundance, the yield total (RYT) for the mixture should be 0.5*Mon A + 0.5*Mon B, where Mon A and Mon B is monoculture yields for species A and B planted at the same overall density, respectively. When this term is >1, over-yielding has occurred, and when it is <1, under-yielding has occurred. Loreau and Hector (2001) modified these measures into a net biodiversity effect (NBE), re-centering RY = 1 to NBE = 0 when no over- or under-yielding is occurring. A positive NBE measure denotes over-yielding and a negative value denotes under-yielding. The net biodiversity effect is composed of two components, the selection effect and the complementarity effect (Loreau and Hector 2001). These terms come from the Price equation in population genetics, which describes how natural 'selection' and other, non-selection processes affect fitness. In ecology, the term 'selection effect' is used interchangeably with the 'sampling effect' by some authors, but it is not exactly the same. The selection effect is the covariance between how a species does in monoculture and its yield in mixture. Species that have high biomass in monoculture and that overyield in mixture have a positive selection effect, and species with low biomass in monoculture but that overyield in mixture have a negative selection effect. The complementarity effect is based on mean yielding behavior after the selection effect has been statistically removed. Thus, it is similar to De Wit's relative yield total, but importantly it is a pure effect that does not include the effects of the selection effect. Fox (2005) developed additional equations that partitions out additional variation from the pure complementarity effect. His approach identifies two types of selection that can be added together and removed from the complementarity effect.

De Wit et al. (1966) went on to study yielding between grass-legume mixtures with *Rhizobium* free legumes vs. *Rhizobium* present legumes. *Rhizobium* are the bacteria that fix N from gaseous into useable plant available forms in the roots of legumes. They found that overyielding only occurred when *Rhizobium* were present (RYT >1). Mixtures with *Rhizobium* free legumes had RYT close to and not significantly different from 1. The legume was using fixed N from the atmosphere, and the grass was using N from the soil, so having both species in roughly equal proportions led to a greater amount of N being available (gaseous + inorganic soil N) to the community as a whole, leading to higher productivity. This early study remains one of the clearest examples of complementary resource use that leads to greater yield in plant mixtures.

The final mechanism that might explain the relationship between diversity and productivity is related to the effects of higher trophic levels and pest outbreaks. This has been much less studied than the other mechanisms (Guerrero-Ramirez et al. 2017). Species in higher trophic levels can moderate the diversity-productivity relationship. Soil pathogens and aboveground pests are more likely to have strongly negative effects in species-poor plots, especially in monoculture plots (Root 1973, Wilsey and Polley 2002). If they reduce productivity only in species-poor plots, then monocultures will be reduced below the diverse mixture levels, leading to a saturating relationship (Schnitzer et al. 2011). Pathogens and root feeding nematodes could also build-up over time, which would lead to an increasing relationship over time between diverse mixtures and monocultures (Guerrero-Ramirez et al. 2017).

Understanding the mechanisms behind the diversity-productivity relationships is critical for applying experimental results to field situations. It is also important to consider how random species loss is in nature, as this is the most common way that species diversity gradients are created in experiments. Grassland managers have a choice on which species to plant or favor with management efforts, and they are not randomly chosen. For example, in developing restoration seed mixes, if species loss is random, and it is unknown if the species being considered are different in their growth rates, then a greater number of species will always be better than a smaller number of species. A greater number of species selected will ensure that the key species will be present in the seed mix due to the sampling effect. However, in some cases, it is well known to managers which species are the most productive (Dickson and Gross 2015), and they often will include these species in all mixes. If the sampling effect was solely responsible for diversity-productivity relationships, then biodiversity-ecosystem function experiments may not be too helpful to restoration because these key (usually dominant) species will probably be in all seed mixes regardless of their diversity. Biofuel plantings are normally planned to either have dominant grass species or a mixture of multiple species. This has been found to be important in biofuel plantings where switchgrass plantings had similar yield to diverse plantings (Dickson and Gross 2015). Fortunately, all biodiversity-ecosystem functioning relationships are not due solely to the sampling effect (Wilsey and Potvin 2000, Loreau and Hector 2001). Alternatively, if the complementarity effect is the sole mechanism behind diversity-productivity relationships, then a diverse mix of functionally different species will be the most productive. Having species that grow at different times of the year (temporal complementarity), or in different rooting zones could enable diverse plots to be more productive than less diverse plantings.

In some cases, it has been found that the effect saturates relatively quickly (Cardinale et al. 2007), even after effect sizes are scaled up. This suggests that effects may be most pronounced when comparing managed monocultures to mixtures rather than mixtures with medium diversity to mixtures with high diversity that may be more common in field situations. Some restoration studies have found small effects of diversity on ecosystem processes because they are comparing medium and high diversity situations, which is the area above the saturation point.

Meta-analyses of biodiversity studies have found support for both the selection and complementarity effect, with the selection effect varying from negative in 43 percent of cases studied by Cardinale et al. (2007), and positive in the rest. A negative selection effect occurs when a low yielding species overyields in mixture. Perhaps most importantly, the complementarity effect tends to become stronger with time in most experiments, and the mechanism often shifts from the selection effect in the first few years to the complementarity effect in later years (Cardinale et al. 2007). This was found experimentally by Tilman et al. (2001) in the Cedar Creek diversity experiment (Figure 6.2). In early years, relationships were fan-shaped and saturating, which supports the selection and sampling effects as partially responsible for relationships. In later years, the relationship became much less fan-shaped and much more linear, providing support for the increasingly important role for complementary resource use. Soil nitrate is highly mobile, and it can leach through the soil column if it is not taken up by vegetation. Plots with a greater number of species prevented nitrate loss, and utilized a

Figure 6.2   The "BigBio" study at Cedar Creek Reserve in Minnesota, USA was established in 1994 to test whether ecosystem process rates are caused by species diversity of the plant community. Plots were seeded with 1, 2, 4, 8, or 16 perennial grassland species in 168 plots that were 9 x 9 m in size. Species were randomly selected from a overall species pool of 18 species (4 species each from the four main functional groups of C_4 grasses, C_3 grasses, Legumes, and forbs, plus 2 species of woody species). Numerous influential publications have resulted from this study, including Tilman et al. (2001) that found that saturating relationships between species richness and productivity became stronger and more linear over time, and Tilman et al. (2006) that found that stability in biomass production over time was greater in diverse plots than in less diverse plots. Photo credit: Jacob Miller, used with permission.

greater proportion of N available to the community (Tilman et al. 2001). Loreau and Hector (2001) found that species-rich plots had a greater complementarity effect than low-richness plots in the Biodepth experiment eight sites across Europe (Figure 6.3). Complementarity effects are associated with having species growing at different times of the year, having different rooting depths (Dimitrakopoulos and Schmid 2004, Dornbush and Wilsey 2010), and having legumes with N-fixing bacteria in their roots and non-fixing species complementing each other. In all of these cases, it is assumed that these differences lead to a greater proportion of resources (e.g., light, water, N, or P) being captured within a growing season.

A recent study (Zuppinger-Dingley et al. 2014) found that the complementarity effect can be greater in the generation following the current one. They collected seeds and tillers from plants in their biodiversity-ecosystem functioning experiment, and then planted them into new plots in monocultures

(a)

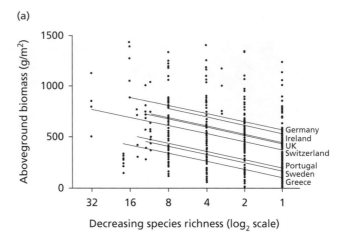

(b)

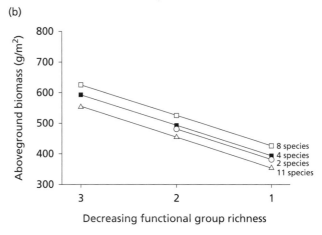

Figure 6.3   Biodepth experiment—The beginning of network ecology? The Biodepth experiment varied species richness at the plot level (x axis) and measured how peak biomass responded (y axis), and this design was repeated in multiple countries throughout Europe. Although previous studies have looked at experimental treatments at multiple sites, for example, competition between *Poa pratensis* and neighboring plants (Reader et al. 1994), the Biodepth experiment was arguably the first large-scale network study. It involved a common experiment distributed across 8 countries in Europe. The study compared plots planted with different numbers of species (a) and functional groups (b). Ecosystem functioning changed as the number of species planted decreased. A companion network included a network of experiments that varied species evenness (Kirwan et al. 2007). From Hector et al. (1999), used with permission.

and mixtures. They found a much higher level of complementarity among the offspring of the original community members. This interesting study suggests that species interactions led to increased complementarity, and that complementarity effects may continue to increase over time across generations.

6.4 Studies with different measures of diversity and different response variables

Other studies have varied species evenness. By varying species evenness, all species are included in plots being compared across diversity levels, which can be used to exclude the sampling effect from consideration (Wilsey and Potvin 2000). Wilsey and Potvin (2000) found that plots planted with higher evenness at higher total ANPP (roots + shoots) than plots with lower evenness. Mulder et al. (2004), and Polley et al. (2003) found a negative relationship and no relationship between evenness and productivity, which suggests that evenness effects can vary with the species being considered. Follow-up studies have found general support for this, with evenness effects varying depending on the species being considered (reviewed by Hillebrand et al. 2008). Variation among studies can partially be explained by considering whether the authors used perennials (Wilsey and Potvin 2000, Mulder et al. 2004) or annuals (Polley et al. 2003) in their study designs.

Other ecosystem processes can also be affected by biodiversity. Studies have been done on all aspects of C and nutrient cycling, but the effects of biodiversity on net primary productivity are over-represented. However, C cycling depends not only on net primary productivity, which is the amount of C that is produced over time, or GPP—total R, but decomposition as well. The C produced by NPP undergoes decomposition and the amount of C that accrues in the system (sometimes called ecosystem production) is a function of NPP—amount decomposed over time. Given that, it is surprising how few studies have been done on decomposition. Those that have been conducted have found that results are somewhat mixed. Species richness of litter is not consistently related to decomposition in herbaceous communities (Wardle et al. 1997, Knops et al. 2001), but is in forest and stream communities (Handa et al. 2014). Evenness can have important effects on litter decomposition, with litter mixes with high evenness having higher decomposition than mixtures with low evenness (Dickson and Wilsey (2009), and Swan et al. 2008, Gessner et al. 2010). Decomposition is lower when there is a poor balance of nutrients and high secondary compounds like phenolics and lignin. Litter mixtures with relatively equal abundances among species may lead to a more balanced diet and lower secondary compounds to decomposing microbes. However, the effects of evenness can be small compared to species composition and landscape position effects (Cornwell et al. 2008, Dickson and Wilsey 2009). In general, most belowground processes that are important in soils are poorly related or idiosyncratically related to species richness (Bardgett 2005).

If decomposition is less effected by diversity than NPP is, then increasing diversity should lead to greater C accumulation in the soil. Few studies have tested this, but Mueller et al. (2013) found increased soil C accumulations in diverse plots. High richness plots at Cedar Creek had a greater proportion of deeply-rooted species, and this was associated with higher soil C compared to plots with fewer species. This suggests that in at least some cases, C can be sequestered in soils by managing for high diversity in grasslands.

Other studies have compared non-random and random extinction. Zavaleta and Hulvey (2004) found that diversity-productivity effects were stronger when using nested species loss compared to random species loss. Larsen et al. (2005) found that bee effects on plant pollination success and dung beetle dung burial were larger with realistic extinction loss than random extinction. In both cases, the larger bees or dung beetles were more efficient in their activities than small bodied species. Large bodied species were also more likely to go extinct than small bodied species, suggesting that random extinction studies are under-estimating the effects of species loss. Smith and Knapp (2003) removed species from plots, and varied the proportion of a dominant grass species, and found that the dominant grass species was more important than richness to productivity (also see McNaughton 1983). Isbell et al. (2008) found that losing tall species led to greater effects on primary productivity than losing short species across diversity levels.

6.5 Multiple ecosystem processes

Finally, it is important to look at multiple ecosystem measures rather than single measures like NPP, and this is an active area of research. If the key species differ between functions, then analyzing a single function in isolation will underestimate the effect of diversity on ecosystem function. Consistent with this, Hector and Bagchi (2007) and Zavaleta et al. (2010) found that relationships between biodiversity and ecosystem functioning were stronger when multiple measures are being considered compared to a single measure. This makes sense, because the species most important to productivity are probably not the most important to decomposition and other processes. Isbell et al. (2011) found that the proportion of species important to ecosystem function increases dramatically as you increase the number of contexts being considered (Figure 6.4). The number of species being important to ecosystem services increased as you included a greater number of years, functions, environmental conditions, or places. Duffy (2009) argued that biodiversity effects on ecosystem processes may be stronger as one considers larger scales in space and time. Scaling up the many small-scale studies to larger spatial and temporal scales remains an important field of research.

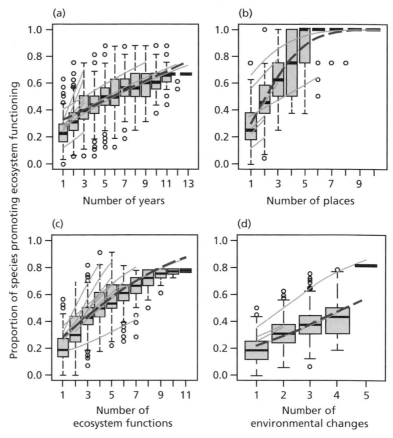

Figure 6.4   The proportion of species that significantly promoted ecosystem functioning increased as one considers the number of years (a), locations (b), ecosystem functions (c), and environmental changes (d). For example, in a given year, a subset of species are important to productivity for example, but a greater subset are important when looking over multiple years (a). A greater proportion of species are important when considering different sites (e.g. slopes and aspects in a hilly area) vs. considering a single site (b). When considering multiple functions, for example productivity and decomposition vs. only productivity, a greater proportion of species are significant predictors (c). The species most important to decomposition are not always the same species that are important to productivity. Finally, a greater number of species promote ecosystem function when multiple environmental conditions are considered (e.g. high vs. low precipitation levels, high vs. ambient CO_2, high vs. low nutrient deposition) compared to a single condition (d). From Isbell et al. (2011), used with permission.

6.6 Biodiversity-stability relationships

There are many interesting questions about how diversity may be related to stability in ecosystem functions over time, and many of the studies have led to controversy in the past. Robert McArthur developed a general prediction for how diversity is related to stability by pointing out that diverse systems will have more pathways for energy to flow through than less diverse systems. Removing some of these pathways should weaken the system as a whole. Early empirical work supported this (McNaughton 1977). For example, McNaughton (1977) found that grassland recovery following buffalo grazing was higher in diverse grassland than in less diverse grassland in Tanzania. However, Robert May's modeling work, also in the 1970s, suggested that food webs with more complexity would be less stable than food webs with less complexity, using species connectivity as a measure of complexity. This apparent contradiction has been partly settled by Tilman et al. (2006), who found that individual species fluctuations, which may appear unstable, can compensate for each other to stabilize things at the community level. Thus, results can depend on what you measure, and how you operationally define stability. Pimm (1984) pointed out that multiple components and definitions of stability exist, and suggested that there may not be a general relationship for all measures and that relationships will vary depending on whether resistance or resilience is being considered.

The early 1990s saw a huge resurgence in this topic with the grassland study of Tilman and Downing (1994). Experimental grassland plots that were fertilized with N went through a drought, which enabled a test of how diversity will influence an ecosystems response to drought. Fertilized plots that had low species richness and diversity had lower resistance (changed more during the drought) and lower recovery following the drought. However, it was pointed out that this early study did not directly vary diversity (Givnish 1994). Givnish (1994) suggested that N additions led to a reduction in root-shoot ratios in fertilized plots, and that low root-shoot ratios were responsible for the low resistance and recovery from drought. This disagreement has been rectified by the large number of studies that have directly manipulated diversity since that time. An analysis of 46 grassland studies from North America and Europe found that species-rich plots had greater resistance to extreme years, either wet or dry using studies that experimentally manipulated diversity (Isbell et al. 2015, Figure 6.5). Recovery (i.e. resilience) was not significantly related to species richness. Cardinale et al. (2013) found that diversity is generally related to both average net primary productivity and average stability across studies, but that these things do not always covary (the averages are different, but the plots with the highest productivity are not necessarily the ones with the highest stability). Studies that have looked for relationships between diversity and stability using unmanipulated (natural) levels of diversity have found a weaker relationship (Grace et al. 2007, Polley et al. 2007, Sasaki and Lauenroth 2011).

Understanding and applying the mechanisms behind diversity-stability relationships are key to applying this work to field situations (Figure 6.6). An understanding of the mechanisms can explain why results are sometimes not consistent across studies. This gets at the question of when diversity is

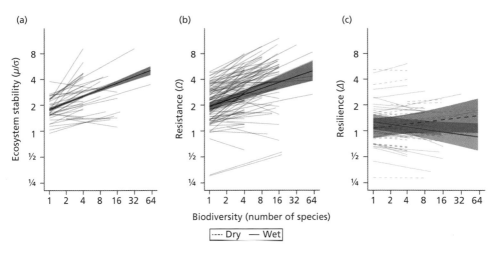

Figure 6.5 Biomass stability, or the consistency of biomass production over time (a), resistance, or the lack of change in biomass during extreme events (b), and resilience, or recovery from extreme events across 46 experiments that varied plant species richness experimentally (Isbell et al. 2015, used with permission).

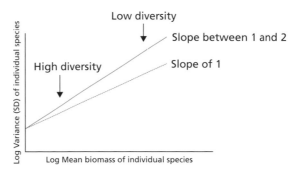

Figure 6.6 The portfolio effect operates to reduce the variability in diverse communities when the slope of standard deviation (SD) plotted against the mean is between 1 and 2. In high diversity communities, the biomass of each species in the community is relatively low ('High diversity'). In low diversity communities, biomass is high ('Low diversity'), leading to a greater SD/mean. The (SD/mean)x100, or coefficient of variation, is the most commonly used measure of variability.

related to stability instead if they are related or not. Variation in results initially came down to whether an individual species or the whole community was being considered. Theory predicts that counteracting fluctuations among species within a community will stabilize abundances and biomass production of the community (McNaughton 1977, Tilman 1999, Loreau and de Mazancourt 2008). Thus, an individual population might be fluctuating wildly in a stable community as long as the fluctuations are balanced by other species in the same trophic level. For example, Tilman et al. (2006)

found that the variation (CV) in biomass production of the plant community was lower but that the variation in populations was higher with increasing species richness. This helped to explain the different predictions made by May's work and more recent diversity-stability work. Loreau and de Mazancourt (2008) developed a measure of species asynchrony to address this question specifically, and their measure has been used in many recent studies on the topic. Asynchronous growth accounted for stability differences and was found to be higher in species-rich plots (Isbell et al. 2009, Hector et al. 2010), in plots not receiving N deposition (Hautier et al. 2014), and in uninvaded native communities compared to invaded exotic communities (Wilsey et al. 2014).

Biodiversity is not the only thing that is important to stability however, and the stability of dominant species can also be important (Lepš 2004). Polley et al. (2007) compared pairs of diverse remnant and less diverse restored prairies, and found no difference in stability in one pair. In the second pair, the less diverse restored prairie had higher stability than the more diverse remnant. This was due to a dominant species (*Schizachyrium scoparium*) that was especially stable, being more abundant in the restored site. There is some support for this mechanism in old field communities of Michigan. Sasaki and Lauenroth (2011) found that the dominant grass *Bouteloua gracilis* is especially stable over time, and that grassland sites with more *Bouteloua* were more stable than more diverse sites with lower dominance by this species. Wilsey et al. (2014) found that there were two ways to achieve stability by comparing diverse native and less diverse exotic dominated grassland plots: either via species asynchrony (in native plots) or by having especially stable dominant species (in exotic plots). The dominant species effect was enough to result in no difference in stability between native and exotic plots even though there was a very large difference in species diversity.

Finally, the portfolio effect can be an important mechanism that leads to stability. The portfolio effect is due to the reduced variance found when individual units are averaged. The mean of the units is generally less variable than the original units. The portfolio effect can be quantified with the general relationship between variance and the mean (Taylor 1961) of individual species on a log-log scale. If the slope is between 1 and 2 (and the portfolio effect is operating), it means that the variance-mean ratio will be larger at higher values than it will be at lower ones. For example, mean biomass of a species in monoculture tends to be greater than biomass per species of each species in a mixture. Based on their relatively low abundance in mixture, the ratios of variance to the mean (CV) will be low in each species in mixture. In monocultures, high abundance will lead to high variance to mean ratios (CV). The portfolio effect is common in biodiversity-ecosystem functioning experiments and is a common mechanism in explaining why variability over time is lower in diverse plots than in less diverse plots (Figure 6.6).

7 Response of Grasslands to Global Change

7.1 Introduction

Global change factors are ecologically-relevant variables that are changing, and that have global impacts. Not all ecological factors are global in scope, and some of the factors that are the most important locally (e.g., changes in fire or grazing intensities) are not considered under the global change umbrella. As a result, there is some disagreement about what is considered to be global change. Most grassland scientists would agree that changes in the atmosphere, biological invasions, N deposition, and land-use change are global change factors.

The atmosphere is globally distributed, and gases diffuse quickly from one location to another, so factors that contribute to directional change in atmospheric composition are considered global change. CO_2 concentrations, temperatures, and N_2O concentrations are rising due to fossil fuel burning (Vitousek 1994). Rainfall is becoming more variable (Figure 7.1). Land-use change is usually considered global because of world-wide increases in human populations. N-deposition, alterations to the N cycle, and biotic homogenization are so widespread that they are usually considered global change (Vitousek 1994). **Biotic homogenization** from the introduction and spread (invasion) of non-native species is especially important in grasslands. In this chapter, I discuss global changes that result from atmospheric changes first, including their effects on phenology and phenological mismatches between plants and animals. Then, I will discuss effects of global changes related to species introductions, including biotic homogenization. Planting and intentionally introducing non-native species into grasslands is widespread and represents a form of land-use change. Reversing this effect to favor native species is a form of restoration, a topic to be covered in Chapter 8.

The Biology of Grasslands. Brian J. Wilsey. Published 2018 by Oxford University Press. © Brian J. Wilsey 2018.
DOI 10.1093/oso/9780198744511.001.0001

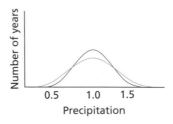

Figure 7.1   Climate models (gray curve) are predicting more extreme years compared to current conditions (black curve). Mean precipitation is standardized as 1.0. Extreme years can be especially wet, causing flooding, or especially dry, causing drought.

7.2 Atmospheric changes

The composition of the atmosphere is no longer in an equilibrium state. Shifting atmospheric conditions are responsible for numerous components of climate change. These climate change components include elevated atmospheric CO_2, increased temperature, N-deposition, and altered precipitation amounts and frequencies. Following, I review impacts of each climate change component on grasslands, beginning with the direct effects of elevated CO_2. Ecological effects of elevated CO_2 has been studied much more intensively than those of other aspects of atmosphere-related climate change (Knapp et al. 2002, Weltzin et al. 2003, de Dios Miranda et al. 2009), although precipitation studies are now catching up.

The concentration of CO_2 in our atmosphere has increased from approximately 270 ppm in the mid-1800s to approximately 400 ppm in the year 2017. It is expected to rise to 550–700 by the end of the twenty-first century. Elevated CO_2 is expected to have direct effects on plants and ecosystems, and indirect effects from associated changes in air temperature and rainfall patterns. Beginning in the 1980s, many CO_2 enrichment studies focused on direct CO_2 effects on grassland species and grassland ecosystems. Higher CO_2 directly influences plants by increasing rates of leaf-level photosynthesis and reducing leaf conductance to water loss. A second wave of studies considered how rising CO_2 might have: 1) indirect effects via their effects on soil water and species composition, and 2) interactions with other global change factors such as N deposition, altered rainfall patterns, and warming. It becomes complicated when looking at interactions among more than three factors, and some factors cancel each other out in unanticipated ways.

7.2.1 Photosynthetic types

Plants with different photosynthetic types differ in their response to climate change. There is one uncommon (CAM) and two common modes of photosynthesis among grassland plant species. CAM, in which stomata are kept closed during the day to prevent water loss, is found in cacti (e.g., prickly pear, *Opuntia* spp.) and will not be discussed in detail here. We focus rather on climate change effects on the much more common C_3 and C_4 photosynthesis types.

Photosynthesis involves converting captured solar energy into chemical energy in the form of ATP and NADPH. These compounds are then used as an energy source to combine CO_2 and Ribulose Bisphosphate (RuBP) into sugars. The first carbon compound produced by C_3 photosynthesis is a three-carbon compound called PGA. The enzyme that catalyzes this reaction is Rubisco. PGA is converted to glucose in a series of reactions and then glucose is used to synthesize other needed compounds. Rubisco makes up about 40 percent of the total N in the leaf and is a major component of plant protein in animal feed. Rubisco is the most common enzyme in nature, and it has interesting properties, as it is a carboxylase as previously described, but it is also an oxygenase, breaking down RuBP and releasing CO_2 during a process called photorespiration. High rates of photorespiration can greatly reduce growth rates, and net losses of C to photorespiration are approximately 40 percent at the current atmospheric CO_2 concentration conditions (Sage 2004). Photorespiration rates are higher under low CO_2–O_2 ratios (i.e., low CO_2 concentrations) and under higher temperatures. Common C_3 species in grasslands are legumes, forbs, and some grasses, especially from the Pooideae clade. They are called 'cool-season' species by some grassland ecologists because they tend to grow when it is cool (spring and fall) seasonally and in colder regions of the globe.

C_4 photosynthesis is responsible for much of the remainder (~23 percent) of global gross primary productivity (Still et al. 2003). The first step of C_4 photosynthesis involves the enzyme PEP carboxylase, and the first compound produced is a 4-carbon sugar. This enzyme is a carboxylase only, and C_4 plants do not have photorespiration. Under high temperatures or low CO_2 concentrations, C_4 plants have higher photosynthetic rates and greater production than C_3 species due to this lack of photorespiration (Figure 7.2). The 4-carbon compound is transported through active transport (requiring energy) to the bundle sheath layers around the leaf veins. There it is released as CO_2 in a concentrated form (i.e., thereby increasing the CO_2:O_2 ratio) to Rubisco. The CO_2 combines with RuBP to form carbohydrates. In C_4 plants, the C_3 pathway is embedded in the bundle sheath, where the O_2 content is low, and Rubisco acts only as a carboxylase preventing photorespiration. Under low CO_2, C_4 species have higher growth rates than C_3 species, and their initial global spread was found to occur when CO_2 was especially low (Cerling et al. 1997, Edwards et al. 2010). However, the advantage that C_4 species have over C_3 species has lessened as CO_2 has risen since the beginning of the industrial revolution. Experimentally increasing the CO_2 concentration or reducing temperature reduces the C_4 advantage over C_3s because the rate of photorespiration (and carbon loss due the oxygenase

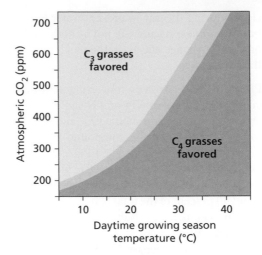

Figure 7.2   High temperatures and low atmospheric CO_2 favors C_4 grasses over C_3 grasses (from Edwards et al. 2010, used with permission). Our current atmospheric CO_2 concentration is 400 ppm and rising. Preindustrial revolution levels were ~270 ppm. Elevated CO_2 will favor C_3 grasses over C_4 grasses but increasing temperatures may reduce this favoritism. Photo: Edwards et al. (2010) used with permission.

reaction) declines in C_3 species, which increases their relative competitive ability. Common C_4 species are grasses such as bluestems, corn, foxtails (*Setaria* spp.), and some sedges. They are called 'warm-season' species by some because they actively grow during summer and in areas with higher annual temperatures.

It is somewhat of a conundrum as to why C_4 plants do not dominate in all ecosystems because of their high potential growth rates. Most forests, and many other ecosystems have species pools that are almost entirely C_3. Why have C_3 dominated systems not been invaded and replaced by C_4 species? The answer lies in a major limitation of the C_4 pathway. The C_4 is costly because it requires more energy per glucose molecule produced than does the C_3 pathway because of the additional step in concentration of CO_2 at the site of the C_3 pathway. As a result, C_4 species are less shade tolerant than C_3 species and are restricted to sunny locations. Because C_4 species are intolerant of shade, their presence in paleontological studies serves as an indicator of open sunny habitats (Cerling et al. 1997, Edwards et al. 2010).

7.2.2 Elevated atmospheric CO_2 studies: effects on plants and ecosystems

Photosynthesis increases when atmospheric CO_2 is experimentally increased around the plant. Based on their physiology, a common prediction is that C_3 species will respond more strongly than C_4 species to CO_2 enrichment. This is true in general, but there is a surprising amount of variation depending on context. Wand et al. (1999) reviewed the literature mostly on studies from greenhouse and growth chamber experiments and found that C_3 grass species responded with a 44 percent increase in growth response to CO_2 enrichment; this value was 33 percent in C_4 grass species. Later research found that field responses were smaller than greenhouse and growth chamber responses, but that the C_3–C_4 difference was still apparent. C_3 grass species responded to CO_2 enrichment with increased tillering (Wand et al. 1999). The growth response in C_4 grasses resulted from improved water status and increased water use efficiency (C fixed/mm precipitation). C_3 grasses consistently had reduced leaf N concentrations under elevated CO_2, whereas C_4 grasses had no significant difference (Wilsey et al. 1997, Wand et al. 1999).

Field studies in grasslands have typically used controlled open and closed top chambers (Figure 7.3), and FACE rings to assess how CO_2 enrichment affects whole ecosystems. Owensby et al. (1993) found higher NPP,

Figure 7.3   Studies on the effects of elevated atmospheric CO_2 have been done in growth chambers, greenhouses, closed (pictured here in Temple, TX) and open-topped chambers, and FACE rings. The site pictured here has many levels of CO_2 applied, from pre-industrial revolution levels, to current levels, to elevated levels expected in the future. Results suggest that CO_2 increase is already having an effect on grassland ecosystems (i.e., that pre- and current CO_2 levels differ significantly). Photo: H. Wayne Polley used with permission.

soil moisture, and forb biomass under elevated CO_2 than under ambient conditions. Morgan et al. (2004) found an increase in ANPP by 41 percent, and a shift from C_4 to C_3 grass dominance in an open-topped chamber study in semiarid rangelands. Interestingly, species composition also shifted from semiarid species towards species found in the more mesic mixed grass prairie system (Morgan et al. 2004, Parton et al. 2007). This shift was similar to the study of Polley et al. (2014), who found that tallgrass prairie species responded more strongly to CO_2 than mixed grass prairie species. Morgan et al. (2011) found that increased water availability under elevated CO_2 nullified the reductions in moisture due to warming of +1.5, +3.0°C (day, night). Elevated CO_2 alone favored C_3 grasses and warming alone favored C_4 grasses. Responses to CO_2 can also be influenced by the number of species present, with greater responses to elevated CO_2 with higher species richness (Reich et al. 2001).

One mechanism behind these growth enhancements to CO_2 is an indirect effect of elevated CO_2 that increases soil moisture (Polley et al. 1997, Morgan et al. 2011. There are strong linkages between the carbon and water cycles in how ecosystems will respond to climate change. Plants have stomata on the undersurfaces of their leaves that regulate gas exchange. When they are open, water can be evaporated (transpiration) into the atmosphere, leading to water stress. Under higher atmospheric CO_2 concentrations, stomata can remain closed more often, which leads to less water lost, and more carbon fixed per unit of water. Carbon fixed per unit of water is called **water use efficiency**, and this can be higher under elevated CO_2. Hence, under elevated CO_2, soils can have higher moisture contents than under ambient CO_2. This increase in soil moisture can lead to increased dominance by mesic species over xeric species, which usually leads to even greater increases in ANPP at elevated CO_2 because mesic species are more productive than xeric species (Polley et al. 2014). This shift in community structure can indirectly enhance growth responses to elevated CO_2 (Polley et al. 2014).

Responses to rising CO_2 concentrations are already occurring. Studies that varied CO_2 from pre-industrial revolution levels to current levels have found that greater responses to CO_2 enrichment effects occurs between subambient (270 ppm) and current levels of CO_2 (400 ppm) than between current and future concentrations (Gill et al. 2002).

7.2.3 Elevated atmospheric CO_2 studies: effects on animals

Under elevated CO_2, plants require less Rubisco, producing leaves and stems with lower protein contents as a result. Leaf N contents are typically lower under elevated CO_2, especially in C_3 species (Wilsey et al. 1997, Wand et al. 1999, Ehleringer et al. 2002). Protein and N are related by the equation N x 6.25 = protein. (The constant 6.25 is based on the average N content of a protein). Lower protein contents under elevated CO_2 are predicted to have negative impacts on herbivores, especially mammalian species, that cannot compensate for protein reductions by feeding longer or by shifting their diet to nutrient rich species. Protein content is positively correlated with digestibility, so lower protein contents mean lower digestibility. Morgan et al. (2004) found reduced digestibility under elevated CO_2 in grasses, which will lower the nutritive value to grass feeders such as bison and cattle. Bison and cattle consume an average of 95 percent and 85 percent grass in their diet, respectively. Ungulates already spend most of their day grazing or processing their food (e.g., chewing their cud). Lower-quality tissues are expected to lower weight gains because they cannot spend additional time feeding to compensate for reduced protein contents (Wilsey et al. 1997, Ehleringer et al. 2002). With domestic livestock, humans can respond to this by fertilizing more with N to compensate for reduced forage quality. Wild animals may be more negatively affected as a result. Grazing mammals could also counter this reduction in forage quality by consuming a greater proportion of species that have inherently higher protein contents (Table 3.1), such as C_3 grasses, forbs, or legumes. This could shift grazing from C_4 grasses to species of these other functional groups (Ehleringer et al. 2002).

Insects may be better able to compensate for nutrient reductions. Reductions in protein content and higher productivity aboveground are expected to alter herbivore feeding. Insects have been found to grow more slowly on plant tissues from elevated CO_2 environments (Lincoln et al. 1993), which makes them more vulnerable to parasitism and predation (Stiling et al. 1999). In scrub oaks, leaf miners had higher mortality rates under elevated CO_2 due to having higher parasitoid rates (Stiling et al. 1999). Lower protein content led to slower development in the leaf miners, which delayed their emergence compared to individuals under ambient CO_2, making them move vulnerable to parasitoid attack. Herbivores may be more vulnerable to predators and parasites in an elevated CO_2 world due to slower development, and this deserves further study.

7.3 Global warming

Atmospheric CO_2 increases are associated with temperature increases because CO_2 and other greenhouse gases like CH_4 and N_2O absorb infra-red radiation reflected from the land surface. Temperature is expected to have significant effects on C cycling, and may lead to positive feedbacks with soil respiration that might lead to CO_2 release to the atmosphere (Raich and Schlesinger 1992). Higher temperature favors plants with C_4 photosynthesis over C_3 species. Along a latitudinal gradient, the proportion of C_4 species declined and the proportion of C_3 species increased moving from south to the north (Teeri and Stowe 1976, Epstein et al. 1997). C_4 species increase in abundance under experimental warming when the warming is applied throughout the growing season (Soussand and Luscher 2007, Morgan et al. 2011). However, there are interesting exceptions to this rule in some grasslands, especially when warming occurs more during the winter and early spring.

The timing of climate changes can be important independently of mean increase effects in grasslands (Alward et al. 1999). T_{min} is the minimum temperature, and this variable is expected to be more impacted by global warming than T_{max}, the maximum temperatures. T_{min} can have unexpected effects that differ from T_{max}. Warming effects are also predicted to be greater nearer the poles than in lower latitudinal areas. As a result, effects of warming in some cases are non-intuitive because they are based more on timing effects than on average effects. Alward et al. (1999) studied decadal responses to T_{min} in short-grass steppes in Colorado, USA. They found that higher T_{min} *decreased* ANPP. Higher T_{min} had this effect because it increased the abundance of exotic and native C_3 forbs that grew in spring before the highly productive and dominant C_4 grasses (mostly *Bouteloua gracilis*) were active. This is an example of a **priority effect**, where warming early in the year favored production of forbs before *B. gracilis* had greened-up. Greater production of forbs reduced the production of the dominant C_4 grass *Bouteloua gracilis* via resource pre-emption, which had the effect of reducing ANPP overall. Priority effects are also important to seedling recruitment and community assembly (Martin and Wilsey 2012, Dickson et al. 2012, Fukami 2015, Wilsey et al. 2015), and response to fertilizer additions (Jarchow and Liebman 2012). ANPP can be reduced overall if the early establishing species has lower than average productivity and has a suppressive effect on the more productive species that grow later. These timing effects can be important in addition to mean effects.

In Arctic tundra warming from 1–3°C led to rapid responses in a coordinated experiment done at 11 locations by Walker et al. (2006). Plant communities responded significantly to warming by the second growing season, and warmed plots had greater heights and cover of shrubs and graminoids, and reduced cover of mosses and lichens. This latter result is likely to affect reindeer and caribou (*Rangifer tarandus*), which relish in eating lichens in winter. Perhaps most importantly, warmed plots had reduced species diversity and evenness, which suggests that biodiversity might be reduced in the future as the climate warms. Their results also suggest that the increase in shrub cover that has occurred in many parts of the arctic in recent years might be due to temperature increases (Walker et al. 2006).

In non-tundra grasslands, the effects of warming are a bit less dramatic. In an annual grassland at Jasper Ridge, CA, Field and colleagues ran a long-term experiment that varied multiple climate change factors. Plots received +300 ppm CO_2 or ambient CO_2, +1°C or ambient temperatures, N of +7 g N m^{-2} yr^{-1} or ambient, and precipitation of +50 percent or ambient. This was an unprecedented study in that it looked at multiple factors simultaneously. Responses were found in functional group composition, phenology, and species diversity. Elevated CO_2 and precipitation had the largest effect on species diversity, whereas, warming led to increases and N additions led to reductions in forbs (Zaveleta et al. 2003). Interestingly, the largest effects on increased forb production was with elevated CO_2, warming and enhanced precipitation combined. This suggests that global change factors will interact.

7.4 Global change effects on phenology

A well-studied aspect of global warming is its effects on phenology. Phenology is the timing of biological events such as green-up, flowering, and senescence. Plant canopies typically green-up, reach a peak and then decline (senesce) towards the end of the growing season (Figure 4.1). A dip in canopy activity is sometimes found during the peak of summer, especially in hot or dry climates (Wilsey et al. 2017). The timing of these events is important to higher trophic levels, and mismatches between timing of green-up/senescence and herbivore, pollinator and microbial activities can be detrimental. Green canopies have higher nutrient contents than dormant canopies, which attracts herbivores (Frank and McNaughton 1992, Rivrud et al. 2016). Animals can track changes in phenology and alter their activities and movements to match peak canopy growth and forage quality. Senescence at the end of the growing season is as important as green-up, especially because it can affect animals that migrate (Fridley 2012, Wilsey et al. 2017). Nevertheless, it has been less studied compared to green-up (Gallinat et al. 2015).

7.5 Warming effects on phenology and phenological mismatches

Phenology is affected by plant photosynthetic mode (C_3 vs. C_4), temperature (Tieszen et al. 1997), atmospheric CO_2 (Cleland et al. 2006), altered rainfall patterns (Prevey and Seasted 2014) and species compositional shifts from native towards exotic species (Wilsey et al. 2017). Current estimates are that green-up occurs 2.5–5 days earlier with each degree C increase in temperature (Wolkovich and Cleland 2011). Warming experiments indicate that warming will induce green-up 1.9–3.3 days earlier for each degree C of warming (Wolkovich et al. 2012). This would equate to a 4.8–8.3 days earlier green-up for the expected 2.5°C increase by the end of the twenty-first century.

Increasing temperatures have led to changes in flowering time. Fitter and Fitter (2002) compared flowering times in 385 British plant species between the warmer 1990s and the cooler period of 1954–1990, and found that 16 percent species flowered significantly earlier. A few (3 percent) flowered significantly later. Spring flowering species were most responsive to temperature, and annuals and insect-pollinated species were more likely to flower earlier than perennials and wind-pollinated species. Changing flowering times have the potential to greatly disrupt pollination, possibly leading to reduced seed set and population sizes.

Earlier green-up will increase the length of the growing season, but only if middle and later stages are not impacted. However, earlier green-up can come at a cost if soil resources such as water are limiting (Steltzer and Post 2009), and dips in canopy activity or reductions in peak normalized difference vegetation index (NDVI) can occur during midseason (Wilsey et al. 2017). Mid-season dips in NDVI could lead to no changes in annual productivity even when green-up is earlier and senescence is later under warming. Midseason declines can be prominent in areas with higher temperatures (Sherry et al. 2007). Sherry et al. (2007) found that experimentally-induced warming caused earlier flowering in early flowering species and later flowering in later flowering species, which caused greater within-season bimodality in flowering in Oklahoma grasslands.

Earlier green-up can also make flowering species more susceptible to late spring frosts. Three flowering species in montane grasslands that had earlier green-up (Inouye 2008) had much greater rates of frost kill in June. Inouye (2008) suggests that this may lead to declines in montane wildflower species. Wilsey et al. (2011) found that exotic grassland species greened up earlier than native grassland species in an experiment in the southern plains, USA. This made them suffer more from a late spring frost than later emerging native species. However, the exotics fully recovered by the end of the growing season.

Perhaps the most critical issue in phenological responses to global change is whether phenological mismatches will occur between plants and pollinating and herbivorous animals. Flowering in plants and emergence in animals uses day-length, temperature, and sometimes rainfall as cues. Mismatches are most likely to happen when the cues for plant and animal activity are different. If higher temperatures cause earlier green-up or later senescence (temperature cue), this will not lead to mismatches if the pollinators or herbivores are also responding to temperature. Plants and animals will shift in concert. However, if animal emergence is related to day-length as it is in some insects, and the timing of plant growth is controlled by temperature or precipitation, then mismatches might occur. This will have important implications to plant-animal interactions.

7.6 Nitrogen deposition

Human production of N exceeds the total amount produced from natural nitrogen fixation and lightning (Vitousek et al. 1997). Feeding the growing human populations will require the continued production and use of N to grow food. N additions can have complex effects on ecosystems, but most commonly they cause declines in species diversity, as explained in Chapter 4. N is often the most limiting element in grassland ecosystems, so it can interact with how grasslands respond to all other global change factors. N limitation can reduce the response to elevated CO_2, and adding N experimentally can result in much larger responses to elevated CO_2 (Reich et al. 2001).

By far the longest running experiment on N fertilization is the Park Grass Experiment in Rothamsted, England (Silvertown et al. 2006). The experiment was installed in 1856 and has run continuously since then. It was set up to test the effects of inorganic fertilizers compared to organic manure, with treatments of nothing or P, K, Mg, Na, and N additions. Additions of fertilizer have reduced species diversity and have increased primary productivity overall. Additions of N as NH_4 led to increases in acidity. These reductions in soil pH resulted in reduced species diversity after a few decades, so a liming treatment was added. Fertilizer + Lime had a greater number of species per plot than fertilizer alone, which provided early support for the importance of soil pH on diversity. Later studies have found evolution in the dominant grass populations in response to the treatments (Silvertown et al. 2006).

7.7 Rainfall amounts and variability

Climate change models predict greater precipitation in the northern latitudes, and lower precipitation in southern latitudes, but smaller-scale regional predictions vary. For example, the Hadley and Canadian models both predict that warming-induced evaporation in the summer will lead to drier soils in the central USA Great Plains, but predict much wetter (Hadley model) and slightly wetter (Canadian model) soils in the southern Great Plains and the southeastern US (US National Assessment Synthesis Team of the US Global Research Program 2000, Weltzin et al. 2003). Precipitation increased by ~10 percent over the twentieth century in the sub-humid southern tallgrass prairie region (Karl and Knight 1998), partially due to increased summer precipitation from tropical storm activity (Allan and Soden 2008). In the central and southwest portions of the US grassland region, climate models predict a greater frequency of drought and reduced soil moisture. These changes in precipitation amounts, especially droughts, will have important impacts on grassland productivity, exotic invasions, and species diversity.

Most processes in grasslands are impacted by water availability (Sala et al. 2012, Huxman et al. 2004, Ponce Campos et al. 2013). The least understood aspect of precipitation effects is how variability affects ecosystems, especially when variability is extreme (e.g., extremely wet or dry years vs. average years; Figure 7.1). Intra- and inter-annual variability in precipitation is projected to increase as climate changes (IPCC 2007). Knapp et al. (2002) found that greater variability in rainfall can affect rare forb species in grasslands. Williams et al. (1998) and Fay et al. (2003) found a smaller effect of rainfall variability on dominant grasses. Fay et al. (2008) varied precipitation amounts and length of time between rainfall events and found that both interacted to impact photosynthesis and ANPP in Kansas grasslands. However, responses to rainfall varied across grassland types: Heisler-White et al. (2009) found that ANPP responses to less frequent, but greater rainfall treatments were negative in a tallgrass prairie site, but were positive in mixed grass prairie and shortgrass steppe. This emphasizes the need to have multiple sites in these types of studies.

Extreme years, whether exceptionally dry or exceptionally wet, could alter establishment success, native-exotic proportions, and the strength of priority effects. This in turn could lead to new combinations of species ('novel' communities) that may persist for long time periods. In the Swiss Alps, extremely dry years led to long-term reductions in grass percent cover, and this reduction persisted for decades after the drought occurred (Stampfli and Zeiter 2004). Albertson and Tomanek (1965) noted similar shifts in abundance of dominant grasses that seemed to be associated with droughts.

7.8 Drought legacies

Droughts can have indirect effects that extend beyond the drought period, termed **drought legacies** (Sala et al. 2012). Drought legacies are especially likely to be important in cases where species composition shifts dramatically as a result of the drought (Figure 7.4). A negative drought legacy is found

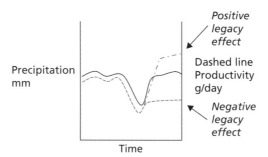

Figure 7.4   Drought legacy effects happen when biomass production no longer tracks precipitation levels after a drought or other extreme event. Dashed line denotes productivity, which can be higher or lower than expected after a drought event (precipitation is solid line). If productivity is reduced below expected values after the drought, the legacy effect is negative, and is it is higher than expected, the legacy effect is positive.

when NPP is below the value expected for a given level of precipitation in the year following the end of the drought, as found by Xu et al. (2017). A positive drought legacy is also possible, especially if unused N builds up during the drought period, which might lead to higher NPP in the year following the drought (Blair 1997).

7.9 Biological invasions and human introductions

Most grasslands are being impacted by biological invasions to various degrees. There are multiple definitions of invasive species, and inferences made from studies can vary depending on the definition used. Richardson et al. (2000) defined **invasive species** as a non-native species that has produced reproductive offspring far from the parent plants. Specifically, >100 m from the parent plants for seedlings, and >6 m for clonal offspring such as roots, rhizomes, and stolons. According to this definition, most non-native species that have developed reproducing populations are 'invasive', regardless of how they impact the environment. A second definition is that an invasive species is a rapidly spreading species. The third and most commonly used definition is a non-native species that is having a negative impact on their environment (Weber 2017), sometimes this is defined as a negative economic impact. Richardson et al. (2000) suggest the terms **'pests'** and **'weeds'** for these species, and **'transformers'** for species having large effects on ecosystem processes. Because of these multiple definitions, lists of invasive and non-invasive species can vary depending on the people assembling them. Here, I will be using Richardson's definition, and I will usually just use non-native species or exotic species to describe non-native species that have spread from their point of introduction. Thus, some invasive species have positive economic impact (e.g., by producing forage), but negative environmental impacts (e.g., if they reduce biodiversity). Many non-native species are used in landscaping and in agriculture, and if they have not spread from the garden or crop field, they are not invasive by definition.

Grasslands are perhaps the biome that is most altered by non-native (exotic) species (Table 7.1). Among the 500 documented invasive plant species that are causing harm to native plants of the world described by Weber (2017), 260 or 52 percent were invading grasslands. Only 6 or 1.2 percent were invading deserts, and 200 or 40 percent were invading forests. The rest were invading wetlands and riverine systems or were invading multiple systems.

Some species have escaped in seed mixes and by other accidental means, but most species were intentionally introduced. **Introduced species** are defined as species that have been transported across major geographical barriers by humans (Richardson et al. 2000). Introduced species were not randomly

Table 7.1 Grassland plant species documented to have negative effects on environment (adapted from Weber 2017). Habitat is P = pasture, G = grassland, S = savanna, and O = old field (human derived grassland), H = heathlands. L Form = life form. The native column identifies which continent the species is native to, and the continent invasive is the continent that it is invading. N/A = not available. See Figure 7.7 for one example.

Habitat	Scientific name	Family	Life Form	Native	Continent Invasive	Commercial uses
G	Acacia baileyana F. Muell.	Leguminosae	Evergreen shrub, tree	Australia	None	Ornamental, erosion control, honey production
G	Acacia cyclops A. Cunn. Ex G. Don	Leguminosae	Evergreen shrub, tree	Australia	Africa, North America	Ornamental, erosion control
G	Acacia longifolia (Andrews) Willd.	Leguminosae	Evergreen shrub, tree	Australia	Africa, Australia, Europe	Ornamental, erosion control
G	Acacia dealbata Link	Leguminosae	Evergreen shrub, tree	Australia	South America, Africa, Europe	Ornamental, erosion control, honey production, shade/shelter, fuelwood
G	Acacia mearnsii De Wild.	Leguminosae	Evergreen tree	Australia	Africa, Europe	Ornamental, erosion control, honey production, shade/shelter, fuelwood
G	Acacia saligna (Labill.) H.L. Wendl.	Leguminosae	Evergreen shrub, tree	Australia	Australia, Africa, Europe	Erosion control, revegetator, fuelwood
G	Acer pseudoplatanus L.	Aceraceae	Deciduous tree	Europe	Europe, Australia	Ornamental, shade/shelter, wood
G	Aegilops triuncialis L.	Poaceae	N/A	Europe and Africa	North America	N/A
G	Agapanthus praecox subsp. Orientalis F.M. Leight.	Amaryllidaceae	Perennial herb	Africa	Australia	Ornamental
G	Agave americana L.	Asparagaceae	Succulent	North America	Australia, Europe, Africa	Ornamental
G	Agave sisalana Perrine	Asparagaceae	Succulent	N/A	Australia, Africa	Ornamental, fibre

	Species	Family	Life form	Native range	Introduced range	Uses
G	Ageratina adenophora (Spreng.) R.M. King & H. Rob.	Compositae	Perennial herb or subshrub	North and South America	Africa, Australia, Asia	Ornamental
G	Agrostis capillaris L.	Poaceae	Perennial herb	Europe and Asia	Australia, North America	Ornamental, erosion control, lawn/turf, fodder
G	Ailanthus altissima (Mill.) Swingle	Simaroubaceae	Deciduous tree	Asia	Australia, North America, Europe	Ornamental, shade/shelter, wood
H	Aira caryophyllea L.	Poaceae	Annual herb	Africa and Europe	Australia	Forage
P	Albizia lebbeck (L.) Benth.	Leguminosae	Deciduous tree	Africa	Africa, North America	Agroforestry, ornamental, revegetator, shade/shelter, forage, soil improver
G	Alliaria petiolata (M Bieb.) Cavara & Grande	Bassicaceae	Biennial herb	Europe, Africa and Asia	North America	Flavoring
G	Allium triquetrum L.	Amaryllidaceae	Perennial herb	Africa and Europe	Australia	Ornamental
G	Alternanthera pungens Kunth	Amaranthaceae	Perennial herb	South America	Africa, Australia	N/A
G	Ambrosia artemisiifolia L.	Compositae	Annual herb	North America	Europe, Australia	N/A
G	Amelanchier spicata (Lam.) K. Koch	Rosaceae	Deciduous shrub	North America	Europe	Ornamental
G	Amorpha fruticosa L.	Leguminosae	Deciduous shrub	North America	Europe	Ornamental
G	Andropogon gayanus Kunth	Poaceae	Perennial herb	Africa	South America, Australia	Forage, cane
G	Andropogon virginicus L.	Poaceae	Perennial herb	North and South America	Australia	Erosion control, ornamental

(continued)

Table 7.1 (Continued)

Habitat	Scientific name	Family	Life Form	Native	Continent Invasive	Commercial uses
G	*Anthoxanthum odoratum* L.	Poaceae	Perennial herb	Europe, Asia and Africa	N/A	Essential oils
P	*Arctotheca calendula* (L.) Levyns	Compositae	Annual to perennial herb	Africa	Europe, North America, Australia	Ornamental
G	*Argemone ochroleuca* Sweet	Papaveraceae	Annual herb	North America	Africa and Australia	N/A
G	*Aristea ecklonii* Baker	Iridaceae	Perennial herb	Africa	Australia	N/A
G	*Arrhenatherum elatius* (L.) P. Beauv. Ex J. Presl & C. Presl	Poaceae	Perennial herb	Africa and Europe and Asia	Australia	Erosion control, forage
G	*Asparagus asparagoides* (L.) Druce	Asparagaceae	Perennial herb	Africa	North American, Australia	Ornamental
G	*Asparagus officinalis* L.	Asparagaceae	Perennial herb	Africa, Europe and Asia	Australia	Vegetable
G	*Asphodelus fistulosus* L.	Xanthorrhoeacea	Perennial herb	Africa and Europe	Australia	Ornamental
G	*Avena fatua* L.	Poaceae	Annual herb	Africa, Europe and Asia	Africa, Australia, North America	Fodder, forage
P	*Berberis darwinii* Hook.	Berberidaceae	Evergreen shrub	South America	Australia	Ornamental
G	*Berberis thunbergii* DC.	Berberidaceae	Deciduous shrub	Asia	Europe and North America	Ornamental
G	*Bidens pilosa* L.	Compositae	Annual herb	North America	Africa, Asia and Australia	Vegetable
G	*Brassica tournefortii* Gouan	Brassicaceae	Annual herb	Africa, Europe/Asia	North America, Australia	N/A
G	*Briza maxima* L.	Poaceae	Annual herb	Africa and Europe	Australia, North America, South America	Ornamental
G	*Bromus inermis* Leyss.	Poaceae	Perennial herb	North America, Europe and Asia	North America	Erosion Control, fodder, forage

G	*Bromus tectorum L.*	Poaceae	Annual herb	Africa, Europe and Asia	North America	Forage, cane
G	*Buddleja davidii Franch.*	Scrophulariaceae	Deciduous shrub	Asia	Europe, Australia, North America	Ornamental
G	*Caealpinia decapetala (Roth) Alston*	Leguminosae	Evergreen shrub, vine	Asia	Australia and Europe	Boundary/barrier, beads
G	*Calluna vulgaris (L.) Hull*	Ericaceae	Evergreen dwarf shrub	Europe and Asia	Australia	Ornamental, erosion control, honey production
P	*Calotropis procera (Aiton) W.T. Aiton*	Apocynaceae	Evergreen shrub, tree	Africa and Asia	North America, South America, Australia	N/A
G	*Carduus acanthoides L.*	Compositae	Annual or biennial herb	Africa, Europe and Asia	North America, South America, Australia	N/A
G	*Carduus nutans L.*	Compositae	Annual or biennial herb	Europe and Asia	North America, South America, Australia	N/A
G	*Carduus pycnocephalus L.*	Compositae	Annual or biennial herb	Africa and Europe and Asia	North America, South America, Australia	N/A
G	*Carrichtera annua (L.) DC.*	Brassicaceae	Annual herb	Africa and Europe	Australia	N/A
G	*Centaurea calcitrapa L.*	Compositae	Annual or short-lived perennial herb	Africa and Europe	North America, Australia	N/A
G	*Centaurea diffusa Lam.*	Compositae	Annual or perennial herb	Europe	North America, South America, Asia	N/A
G	*Centaurea solstitialis L.*	Compositae	Annual herb	Africa, Europe and Asia	North America, South America, Africa, Europe	N/A
G	*Centaurea stoebe subsp. Micranthos L.*	Compositae	Biennial or perrennial herb	Europe	North America	N/A

(continued)

Table 7.1 (Continued)

Habitat	Scientific name	Family	Life Form	Native	Continent Invasive	Commercial uses
G	*Chamaecytisus prolifer (L.f) Link*	Leguminosae	Evergreen shrub	N/A	Australia	Ornamental, fodder
G	*Chasmanthe floribunda (Salisb.) N.E. Br.*	Iridaceae	Perennial herb	Africa	Australia	Ornamental
G	*Chenopodium album L.*	Chenopodiaceae	Annual herb	Africa and Europe	Africa, Australia	Food
G	*Chloris virgata Sw.*	Poaceae	Annual herb	North America, South America, Africa and Asia	North America, Australia	Revegetator, fodder, forage
G	*Chromolaena odorata (L.) R.M. King & H. Rob.*	Compositae	Evergreen shrub	North and South America	North America, Africa, Australia, Asia	Soil improver
G	*Chyrsanthemoides monilifera (L.) Norl.*	Compositae	Evergreen shrub	Africa	Australia	Ornamental
P	*Cinnamomum camphora (L.) J. Presl*	Lauraceae	Evergreen tree	Asia	North America, Africa, Australia	Essential oils, fibre, wood
G	*Cirsium arvense (L.) Scop.*	Compositae	Perennial herb	Europe and Asia	North America, Australia	N/A
G	*Cirsium vulgare (Savi) Ten.*	Compositae	Biennial herb	Africa, Europe and Asia	North America, Africa, Australia	Ornamental
P	*Clidemia hirta (L.) D. Don*	Melastomataceae	Evergreen shrub	North and South America	Africa, Asia, Australia	Ornamental, fruit
G	*Conicosia pugioniformis (L.) N.E. Br.*	Aizoaceae	Perennial succulent	Africa	North America	Ornamental
G	*Conium maculatum L.*	Apiaceae	Biennial herb	Africa, Europe and Asia	North America, Australia	Ornamental
G	*Cortaderia jubata (Lemoine ex Carriere) Stapf*	Poaceae	Perennial herb	South America	North America, Africa, Australia	Ornamental, erosion control

G	Cortaderia selloana (Schult. & Shult. F.) Asch. & Graebn.	Poaceae	Perennial herb	South America	North America, Africa, Europe, Australia	Ornamental, erosion control, fodder
G	Cotoneaster glaucophyllus Franch.	Rosaceae	Semi-evergreen shrub	Asia	Australia	Ornamental
G	Cotoneaster pannosus Franch.	Rosaceae	Semi-evergreen shrub	Asia	North America, Africa, Australia	Ornamental
G	Crataegus monogynza Jacq.	Rosaceae	Deciduous shrub or small tree	Europe and Africa	North America, Australia	Honey production, ornamental
G	Cryptostegia grandiflora R. Br.	Apocynaceae	Evergreen liana	Africa	North America, Australia	Ornamental
G	Cylindropuntia imbricata (Haw.) F.M. Knuth	Cactaceae	Succulent tree	North America	Africa, Australia	Ornamental
G	Cynara cardunculus L.	Compositae	Perennial herb	Africa and Europe	South America, North America, Australia	Ornamental, food
G	Cynodon dactylon (L.) Pers.	Poaceae	Perennial herb	Africa	North America, South America, Europe, Asia, Australia	Erosion control, lawn/turf, fodder, forage
G	Cyperus rotundus L.	Cyperaceae	Perennial herb	Africa, Europe and Asia	North America, Asia	Erosion control, vegetable, forage
G	Cytisus multiflorus (L'Her.) Sweet	Leguminosae	Evergreen shrub	Europe	Australia	Ornamental
G	Cytisus scoparius (L.) Link	Leguminosae	Deciduous shrub	Europe	North America, South America, Africa, Europe, Asia, Australia	Erosion control, ornamental
G	Dactylis glomerata L.	Poaceae	Perennial herb	Africa, Europe and Asia	Australia	Ornamental, fodder, forage

(continued)

Table 7.1 (Continued)

Habitat	Scientific name	Family	Life Form	Native	Continent Invasive	Commercial uses
G	Datura stramonium L.	Solanaceae	Annual herb	North America	Africa, Europe, (small part of) North America	N/A
G	Delairea odorata Lem.	Compositae	Evergreen perennial vine	Africa	North America, Europe, Africa	Ornamental
P	Dichrostachys cinerea (L.) Wight & Arn.	Leguminosae	Shrub or tree	South America, Africa and Asia	Africa, North America	Soil improver, wood
G	Dipogon lignosus (L.) Verdc.	Leguminosae	Evergreen liana	Africa	Australia	Ornamental, soil improver, food
G	Dipsacus fullonum L.	Caprifoliaceae	Biennial herb	Africa and Europe	Australia, North America	Ornamental, honey production
G	Dipsacus laciniatus L.	Caprifoliaceae	Biennial herb	Europe and Asia	North America	Ornamental
G	Echium plantagineum L.	Boraginaceae	Annual or biennial herb	Africa and Europe	North America, South America, Africa, Australia	Ornamental, honey production
G	Ehrharta calycina Sm.	Poaceae	Perennial herb	Africa	Australia, North America	N/A
G	Ehrharta erecta Lam.	Poaceae	Perennial herb	Africa	North America, Australia	N/A
G	Elaeagnus umbellata Thunb.	Elaeagnaceae	Deciduous shrub, tree	Europe and Asia	North America	Ornamental
G	Elymus repens (L.) Gould subsp. Repens	Poaceae	Perennial herb	Africa, Europe and Asia	North America	Fodder, forage
G	Eragrostis curvula (Schrad.) Nees	Poaceae	Perennial herb	Africa	North America, Asia, Australia	Erosion control, revegetator, fodder, forage
G	Eragrostis lehmanniana Nees	Poaceae	Perennial herb	Africa	North America	Erosion control, revegetator, forage
G	Erica lusitanica Rudolphi	Ericaceae	Evergreen shrub	Europe	Australia	Ornamental
G	Erodium botrys (Cav.) Bertol.	Geraniaceae	Annual herb	Africa and Europe	Australia, North America	N/A

G	*Erodium cicutarium* (L.) L'Her.	Geraciaceae	Annual herb	Africa, Europe and Asia	North America, Australia	Fodder
G	*Eucalyptus cladocalyx f. Muell.*	Myrtaceae	Evergreen tree	Australia	Africa, Australia	Ornamental, wood
G	*Eucalyptus diversicolor F. Muell.*	Myrtaceae	Evergreen tree	Australia	Africa	Honey production, wood
G	*Eucalyptus globulus Labill.*	Myrtaceae	Evergreen tree	Australia	North America, Europe, Africa, Australia	Flavoring, honey production, ornamental, essential oils, fuelwood, wood
G	*Euphorbia esula L.*	Euphorbiaceae	Perennial herb	Europe and Asia	North America	N/A
G	*Euryops multifidus (Thunb.) DC.*	Compositae	Deciduous shrub	Africa	North America	N/A
G	*Festuca arundinaceae Schreb.*	Poaceae	Perennial herb	Africa, Europe and Asia	North America, Australia	Erosion control, lawn/turf, ornamental, forage
G	*Foeniculum vulgare Mill.*	Apiaceae	Perennial herb	Africa, Europe and Asia	North America, South America, Australia	N/A
G	*Fraxinus angustifolia Vahl*	Oleaceae	Deciduous shrub, tree	Africa, Europe and Asia	Australia	Ornamental shade/shelter
G	*Genista linifolia L.*	Leguminosae	Evergreen shrub	Africa and Europe	Australia	Ornamental
G	*Genista monspessulana (L.) L.A.S. Johnson*	Leguminosae	Evergreen shrub	Africa and Europe	North America, South America, Africa, Australia	Ornamental
G	*Gladiolus undulatus L.*	Iridaceae	Perennial herb	Africa	Australia	N/A
G	*Gleditsia triacanthos L.*	Leguminosae	Deciduous tree	North America	North America, South America, Africa, Australia	Erosion control, ornamental, shade/shelter, wood

(continued)

Table 7.1 (Continued)

Habitat	Scientific name	Family	Life Form	Native	Continent Invasive	Commercial uses
G	*Gunnera tinctoria (Molina) Mirb.*	Gunneraceae	Perennial herb	South America	Africa, Europe, Australia	Ornamental
G	*Hakea salicifolia (Vent.) B.L. Burtt*	Proteaceae	Evergreen shrub, tree	Australia	Europe, Australia	Ornamental, shade/ shelter, soil improver
G	*Hakea sericea Schrad. & J.C. Wendl.*	Proteaceae	Evergreen shrub	Australia	Africa, Europe, Australia	Erosion control, shade/ shelter, soil improver
G	*Heracleum mantegazzianum Sommier & Levier*	Apiaceae	Perennial herb	Europe (very small area)	North America, Europe, Australia	Ornamental
G	*Hesperis matronalis L.*	Brassicaceae	Perennial herb	Europe and Asia	North America	Oil/fat
G	*Holcus lanatus L.*	Poaceae	Perennial herb	Africa and Europe	North America, South America, Europe, Asia, Australia	Erosion control
G	*Hordeum murinum L.*	Poaceae	Annual herb	Africa, Europe and Asia	North America	N/A
G	*Hyparrhenia rufa (Nees) Stapf*	Poaceae	Perennial herb	Africa	North America, South America, Australia	Fodder
G	*Hypericum perforatum L.*	Hypericaceae	Perennial herb	Africa, Europe and Asia	North America, Africa, Australia	Essential oils
G	*Hypochaeris radicata*	Compositae	Perennial herb	Africa and Europe	North America, South America, Asia, Australia	N/A
G	*Imperata clindrica (L.) P. Beauv.*	Poaceae	Perennial herb	Africa, Europe, Asia and Australia	North America, South America, Australia	Erosion control, ornamental
G	*Jacaranda mimosifolia D. Don*	Bignoniaceae	Deciduous tree	South America	Africa, Australia	Ornamental, wood
G	*Juncus acutus L.*	Juncaceae	Perennial herb	North America, South America, Africa, Europe and Asia	Australia	N/A

P	*Juncus articulatus L.*	Juncaceae	Perennial herb	North America, Africa, Europe and Asia	Australia	N/A
G	*Juncus effusus L.*	Juncaceae	Perennial herb	North America, South America, Africa, Europe and Asia	Australia	Ornamental
G	*Lagurus ovatus L.*	Poaceae	Annual herb	Africa and Europe	Australia	N/A
P	*Lantana camara L.*	Verbanaceae	Evergreen shrub	north America, South America and Africa	Africa, North America, South America, Asia, Australia	N/A
G	*Lathyrus tingitanus L.*	Leguminosae	Annual herb	Africa and Europe	Australia	Ornamental
G	*Lavandula stoechas L.*	Lamiaceae	Evergreen shrub	Africa and Europe	Australia	Flavoring, ornamental, honey production
G	*Lepidium draba L.*	Brassicaceae	Perennial herb	Africa, Europe and Asia	North America, South America, Africa, Europe, Australia	Vegetable
G	*Lepidium latifolium L.*	Brassicaceae	Perennial herb	Africa, Europe and Asia	North America	Flavoring, vegetable
G	*Leptospermum laevigatum (Gaertn.) F. Muell.*	Myrtaceae	Evergreen shrub, tree	Australia	Africa, Australia	Erosion control, ornamental, shade/shelter
G	*Lespedeza cuneata (Dum. Cours.) G. Don*	Leguminosae	Perennial herb or subshrub	Australia and Asia	North America	on control, fodder, forage
G	*Ligustrum lucidum W.T. Aiton*	Oleaceae	Semi-evergreen shrub or small tree	Asia	North America, South America, Africa, Australia	Ornamental
G	*Ligustrum vulgare L.*	Oleaceae	Evergreen shrub	Africa and Europe	North America, Africa, South America, Australia	Boundary/barrier/support, ornamental

(continued)

Table 7.1 (Continued)

Habitat	Scientific name	Family	Life Form	Native	Continent Invasive	Commercial uses
G	*Lonicera maackii (Rupr.) Maxim.*	Caprifoliaceae	Deciduous shrub	Asia	North America	Ornamental
G	*Lotus uliginosus Schkuhr*	Leguminosae	Perennial herb	Africa and Europe	Australia	Forage
G	*Lupinus polyphyllus Lindl.*	Leguminosae	Perennial herb	North America	Australia, Europe	Ornamental
P	*Lycium ferocissimum Miers*	Solanaceae	Deciduous or evergreen shrub	Africa	Australia	Ornamental, shade/ shelter
G	*Marrubium vulgare L.*	Lamiaceae	Perennial herb	Africa, Europe and Asia	North America, Australia	Flavoring, honey production, ornamental
G	*Megathyrsus maximus (Jacq.) B.K. Simon & S.W.L. Jacobs*	Poaceae	Perennial herb	Africa	North America, South America, Australia, Asia	Fodder, forage
G	*Meddia azedarach L.*	Meliaceae	Deciduous tree	Asia	North America, South America, Africa, Australia	Ornamental, wood
G	*Melilotus officinalis (L.) Lam.*	Leguminosae	Annual or biennial herb	Africa, Europe and Asia	North America	Flavoring, honey production, soil improver, fodder, forage
G	*Melilotus indicus (L.) All.*	Leguminosae	Annual herb	Africa, Europe and Asia	North America	Honey production, erosion control, soil improver, forage
G	*Melinis minutiflora P. Beauv.*	Poaceae	Perennial herb	Africa	North America, south America, Australia	Erosion control, forage
G	*Mentha pulegium L.*	Lamiaceae	Perennial herb	Africa, Europe and Asia	North America, south America, Australia	Honey production, ornamental, essential oils
G	*Mesembryanthemum crystallinum L.*	Aizoaceae	Annual or biennial herb	Africa and Europe	Australia, North America	Ornamental, vegetable

G	*Miconia calvescens DC.*	Melastomataceae	Evergreen shrub, tree	North America and South America	Australia	Ornamental
G	*Moraea flaccida (Sweet) Steud.*	Iridaceae	Perennial herb	Africa	Australia	Ornamental
G	*Morella faya (Aiton) Wilbur*	Myricaceae	Evergreen shrub, tree	N/A	N/A	Ornamental
G	*Nassella tenuissima (Trin.) Barkworth*	Poaceae	Perennial herb	North and South America	Africa, Australia	N/A
G	*Nassella trichotoma (Nees) Hack. Ex Arechav.*	Poaceae	Perennial herb	South America	Africa, Australia	N/A
G	*Nicotiana glauca Graham*	Solanaceae	Evergreen shrub, tree	South America0	Africa, Australia	Pollution control
G	*Olea europaea L.*	Oleaceae	Evergreen tree	Africa and Europe	Australia	Fruits, ornamental
G	*Opuntia aurantiaca Lindl.*	Cactaceae	Perennial succulent	South America	Africa, Australia	N/A
G	*Opuntia ficus-indica (L.) Mill.*	Cactaceae	Succulent perennial	North America	Africa, Europe, Australia	Ornamental, fruit, vegetable, fodder
G	*Opuntia stricta (haw.) Haw.*	Cactaceae	Succulent perennial	North and South America	Australia, Africa	Ornamental
G	*Oxalis pes-caprae L.*	Oxalidaceae	Perennial herb	Africa	North America, Europe, Australia	Honey production, ornamental
G	*Oxalis purpurea L.*	Oxalidaceae	Perennial herb	Africa	Australia	Ornamental
G	*Panicum repens L.*	Poaceae	Perennial herb	Africa, Europe and Asia	North America	Erosion control
G	*Paraserianthes lophantha (Willd.) I.C. Nielson*	Leguminosae	Evergreen shrub, tree	Australia and Asia	Europe, Africa, Australia	Ornamental
G	*Parthenium hysterophorus L.*	Compositae	Annual herb	North and South America	Southern Africa, Southern Asia, Northern Australia	N/A

(continued)

Table 7.1 (Continued)

Habitat	Scientific name	Family	Life Form	Native	Continent Invasive	Commercial uses
G	Paspalum conjugatum P.J. Bergius	Poaceae	Perennial herb	North and South America	Australia	Forage
G	Paspalum dilatatum Poir.	Poaceae	Perennial herb	South America	Africa, Europe	Erosion control, fodder, forage
G	Pastinaca sativa L.	Apiaceae	Biennial or short-lived perennial	South America	Southern Africa, Southern Australia	N/A
G	Pennisetum ciliare (L.) Link	Poaceae	Perennial herb	Africa and Southern Europe	Australia, North America	Erosion control, revegetator, forage
G	Pennisetum clandestinum Hochst. Ex Chiov.	Poaceae	Perennial herb	Central Africa	California (North America), Southern Australia	Erosion control, lawn/turf, ornamental, forage
G	Pennisetum macrourum Trin.	Poaceae	Perennial herb	Africa	Australia	Ornamental
P	Pennisetum polystachion (L.) Schult.	Poaceae	Annual or perennial herb	Africa	Australia	Fodder
G	Pennisetum purpureum Schumach.	Poaceae	Perennial herb	Africa	Southern Africa	Erosion control, ornamental, fodder, forage
G	Pennisetum setaceum (Forssk.) Chiov.	Poaceae	Perennial herb	Africa	Southern Africa, North America	Ornamental
G	Pereskia aculeata Mill.	Cactaceae	Succulent perennial	South America	Southern Africa	Fruit, ornamental, vegetable
P	Persicaria perfoliata (L.) H. Gross	Polygonaceae	Annual vine	Asia	North America	Ornamental
G	Phalaris aquatica L.	Poaceae	Perennial herb	Northern Africa and Southern Europe	Australia, North America (California)	Fodder, forage, revegetating
G	Phleum pratense L.	Poaceae	Perennial herb	Southern Africa, Europe and Asia	Small parts of Australia, small parts of North America	Erosion control, fodder, forage, soil improver

G	*Pilosella officinarum Vaill.*	Compositae	Perennial herb	Europe	Small parts of Australia, small parts South America	N/A
G	*Pinus banksiana Lamb.*	Pinaceae	Evergreen tree	North America	Small parts of Australia	Ornamental, wood
G	*Pinus contorta Douglas ex Loudon*	Pinaceae	Evergreen tree	North America	South America, Australia	Wood, erosion control
G	*Pinus nigra j.F Arnold*	Pinaceae	Evergreen tree	Northern Africa and Southern Europe	North America, Europe, Australia	Erosion control, ornamental, wood
G	*Pinus patula Schltdl. & Cham.*	Pinaceae	Evergreen tree	North America	Africa	Ornamental, wood
G	*Pinus pinaster Aiton*	Pinaceae	Evergreen tree	Africa and Europe	South America, Southern Africa, Australia	Erosion control, shade/shelter, wood
G	*Pinus radiata D. Don*	Pinaceae	Evergreen tree	North America	Africa, Australia	Erosion control, ornamental, gum/resin, wood
G	*Piper aduncum L.*	Piperaceae	Shrub or small tree	South and North America	Australia	N/A
G	*Pittosporum undulatum Vent.*	Pittosporaceae	Evergreen shrub, tree	Australia	Southern Africa, Australia, a small part of Europe	Ornamental, essential oils, wood
G	*Plantago coronopus L.*	Plantaginaceae	Annual to perennial herb	Northern Africa and Southern Europe	Australia	N/A
G	*Plantago lanceolata L.*	Plantaginaceae	Perennial herb	Northern Africa and Europe	Australia	Fodder
G	*Poa pratensis L.*	Poaceae	Perennial herb	North America, Africa, Europe and Asia	North America, Australia	
p	*Polygonum aviculare L.*	Polygonaceae	Annual herb	Europe	Southern Africa, Australia	N/A
G	*Populus alba L.*	Salicaceae	Deciduous tree	Africa and Europe	North America, Africa, Australia	Ornamental, shade/shelter, tannin, wood

(continued)

Table 7.1 (Continued)

Habitat	Scientific name	Family	Life Form	Native	Continent Invasive	Commercial uses
G	*Prosopis glandulosa Torr.*	Leguminosae	Deciduous shrub, tree	North America	Africa, Australia	honey production, forage, shelter, erosion control
G	*Prosopis glandulosa Torr.*	Leguminosae	Deciduous shrub, tree	North America	Africa, Australia	honey production, forage, shelter, erosion control
G	*Prunus serotina Ehrh.*	Rosaceae	Deciduous tree	North America	Europe, Australia	Ornamental, wood, shelter
P	*Psidium cattleianum Sabine*	Myrtaceae	Evergreen shrub, tree	South America	North America (Florida), Africa, Australia	Fruit
G	*Psidium guajava L.*	Myrtaceae	Evergreen shrub, tree	South America	Africa, Australia, North America (Florida)	Fruit, wood
G	*Pyracantha angustifolia (Franch.) C.K. Schneid.*	Rosaceae	Evergreen shrub	Asia	North America (California), South America Southern Africa, Australia	Ornamental
G	*Pyracantha crenulata (D. Don) M. Roem.*	Rosaceae	Evergreen shrub	Asia	North America (California), Southern Africa, Australia	Ornamental
G	*Ranunculus repens L.*	Ranunculaceae	Perennial herb	Africa, Europe and Asia	Australia	Ornamental
G	*Reynoutria japonica Houtt.*	Polygonaceae	Perennial herb	Asia	North America, Europe, Australia	Ornamental, vegetable, fodder
G	*Rhamnus alaternus L.*	Rhamnaceae	Evergreen shrub	Africa and Europe	Australia	Ornamental
G	*Rhododendron ponticum L.*	Ericaceae	Evergreen shrub, tree	Europe/Asia	Europe	Ornamental
P	*Rhodomyrtus tomentosa (Aiton) Hassk.*	Myrtaceae	Evergreen tree	Asia	North America (Florida)	Ornamental
G	*Ricinus communis L.*	Euphorbiaceae	Perennial herb to evergreen shrub	Africa	Southern Africa, Australia, North America (California, Florida)	Flavoring, ornamental

				North America	Southern Africa, Australia, Europe and North America (California)	Honey production, erosion control, ornamental, revegetator, shade/shelter, soil improver, wood
G	*Robinia pseudoacacia L.*	Leguminosae	Deciduous tree	North America	Southern Africa, Australia, Europe and North America (California)	Honey production, erosion control, ornamental, revegetator, shade/shelter, soil improver, wood
G	*Romulea rosea (L.) Eckl.*	Iridaceae	Perennial herb	Africa	Australia	Ornamental
G	*Rorippa palustris (L.) Besser*	Brassicaceae	Annual to perennial herb	North America, Parts of Africa, Europe and Asia	Small part of Australia	N/A
G	*Rosa canina L.*	Rosaceae	Deciduous shrub	Africa, Europe and Asia	small part of Australia	Ornamental
G	*Rosa multiflora Thnb.*	Rosaceae	Deciduous shrub	Asia	Southern Africa, North America	Ornamental, erosion control
G	*Rosa rubiginosa L.*	Rosaceae	Deciduous shrub	Europe and Asia	South America, South Africa, Australia	Ornamental
G	*Rosa rugosa Thunb.*	Rosaceae	Deciduous shrub	East Asia	Europe	Ornamental, essential oils, erosion control
G	*Rubus argutus Link*	Rosaceae	Deciduous shrub	North America	Australia	N/A
G	*Rubus cuneifolius Pursh*	Rosaceae	Deciduous shrub	North America	Africa	Flavoring
P	*Rubus ellipticus Sm.*	Rosaceae	Evergreen shrub	Asia	Small part of Africa	Ornamental
G	*Rubus fruitcosus auct. Agg.*	Rosaceae	Deciduous shrub	Europe	Southern Africa, Australia	N/A
G	*Rumex acetosella L.*	Polygonaceae	Perennial herb	Europe and Asia	Small part of Australia, parts of North America	N/A
G	*Rumex crispus L.*	Polygonaceae	Perennial herb	Africa, Europe and Asia	Australia, parts of North America	N/A
G	*Sagina procumbens L.*	Caryophyllaceae	Perennial herb	Northern Africa, Europe and Asia	N/A	Ornamental

(continued)

Table 7.1 (Continued)

Habitat	Scientific name	Family	Life Form	Native	Continent Invasive	Commercial uses
G	*Salpichroa origanifolia (Lam.) Baill.*	Solanaceae	Perennial herb to shrub	South America	Australia	Ornamental, wood
G	*Salsola kali l.*	Amaranthaceae	Annual herb	Europe	North America (California), southern Africa	N/A
G	*Schinus molle L.*	Anacardiaceae	Evergreen tree	South America	North America (California), Southern Africa, Australia	Flavoring, agroforestry, ornamental, wood
G	*Schinus terebinthifolia Raddi*	Anacardiaceae	Evergreen shrub, tree	South America	North America, Southern Africa, Australia	Ornamental
G	*Schismus arabicus Nees*	Poaceae	Annual herb	Africa, Europe and Asia	North America (California)	Forage
G	*Schizachyrium condensatum (Kunth) Nees*	Poaceae	Perennial herb	South America	N/A?	N/A
G	*Schkuhria pinnata (Lam.) Kuntze ex Thell.*	Compositae	Annual herb	North and South America	Southern Africa	N/A
G	*Securigera varia (l) Lassen*	Leguminosae	Perennial herb	Europe and Asia	Parts of North America	Erosion control, ornamental, fodder, forage
G	*Senecio angulatus l. f.*	Compositae	Perennial herb	Southern Africa	Australia	Ornamental
G	*Senecio inaequidens DC.*	Compositae	Perennial herb, subshrub	Southern Africa	Europe	N/A
G	*Senna alata (l.) Roxbl*	Leguminosae	Evergreen shrub, tree	North America	Australia	Erosion control, ornamental
G	*Senna obtusifolia (L.) H.S. Irwin & Barneby*	Leguminosae	Annual to perennial heb	North and South America	Australia	N/A
G	*Sesbania punicea (Cav.) Benth.*	Leguminosae	Deciduous shrub, tree	North America	Southern Africa, North America (Florida)	ornamental
G	*Silybum marianum (l.) Gaertn.*	Compositae	Annual or biennial herb	Northern Africa, Europe and Asia	Australia, North, South America	Flavoring, honey production

	Species	Family	Life form	Native range	Introduced range	Use
G	*Solanum linnaeanum Hepper & P.M.L. Jaeger*	Solanaceae	Evergreen Shrub	Southern Africa	Australia, North America (Florida)	N/A
G	*Solanum mauritianum Scop.*	Solanaceae	Evergreen shrub, tree	South America	Southern Africa, Australia	Ornamental
G	*Solanum nigrum L.*	Solanaceae	Annual or biennial herb	Northern Africa, Europe and Asia	Australia	Fruit
G	*Solanum viarum Dunal*	Solanaceae	Perennial herb	South America	North America (Florida)	N/A
G	*Solidago canadensis L.*	Compositae	Perennial herb	North America	Australia and parts of Europe, Asia	Honey production, ornamental
G	*Solidago gigantea Aiton*	Compositae	Perennial herb	North America	Europe	Honey production, ornamental
G	*Sonchus arvensis L.*	Compositae	Perennial herb	Europe and Asia	Parts of North America	N/A
P	*Sorghum halepense (L.) Pers.*	Poaceae	N/A	Northern Africa, Europe and Asia	North America, Australia	Forage
G	*Sparaxis bulbifera (L.) Ker Gawl.*	Iridaceae	Perennial herb	Southern Africa	Australia	Ornamental
G	*Sporobolus indicus (L.) R. Br.*	Poaceae	Perennial herb	North America, South America, Northern Africa and Asia	Australia, North America (Florida)	N/A
P	*Stachytarpheta cayennensis (h.) Vahl*	Verbenaceae	Perennial herb, subshrub	South America	Australia, small part of Africa	N/A
G	*Stenotaphrum secundatum (Walter) Kuntze*	Poaceae	Perennial herb	Parts of Africa, part of Australia, North and South America	Australia, small part of Europe	Erosion control, lawn/turf
P	*Syzygium jambox (L.) Alston*	Myrtaceae	Evergreen or semi-deciduous tree	Parts of Asia	Australia	Fruit

(continued)

Table 7.1 (Continued)

Habitat	Scientific name	Family	Life Form	Native	Continent Invasive	Commercial uses
G	*Taeniatherum caput-medusae* (L.) Nevski	Poaceae	Annual herb	Northern Africa and Europe/Asia	North America	N/A
G	*Tagetes minuta L.*	Compositae	Annual herb	South America	Southern Africa, Australia, North America (California)	Erosion control, ornamental, revegetator, shade/shelter, fuel-wood, wood
P	*Thunbergia grandiflra* Roxb.	Acanthaceae	Evergreen Climber	Asia	Australia	Ornamental
P	*Tibouchina herbaceae* (DC.) Cogn.	Melastomataceae	Perennial herb, subshrub	South America	N/A	N/A
G	*Triadica sebifera* (L.) Small	Euphorbiaceae	Deciduous tree	Asia	Australia, small parts of North America	Ornamental, shade/shelter, tannin, wood
G	*Trifolium repens L.*	Leguminosae	Perennial herb	Northern Africa, Europe and Asia	Australia	Soil improver, forage
G	*Trifolium subterraneum L.*	Leguminosae	Annual herb	Northern Africa, Europe and Asia	Australia	Erosion control, soil improver, forage
P	*Trubina corymbosa* (L.) Raf.	Convolvulaceae	Perennial vine	North and South America	Australia	Ornamental
G	*Ulex europaeus L.*	Leguminosae	Evergreen shrub	Europe	North America, Asia, Australia, Southern Africa	Boundary/barrier/support, erosion control, ornamental, revegetator
G	*Ulmus pumila L.*	Ulmaceae	Deciduous tree	Asia	Parts of North America	Ornamental, revegetator, shelter
G	*Urena lobata L.*	Malvaceae	Perennial herb or subshrub	South America and Asia	Australia, North America (Florida)	Essential oils, fibre
G	*Urena lobata L.*	Malvaceae	Perennial herb or subshrub	South America and Asia	Australia, North America (Florida)	Essential oils, fibre

H	*Vellerophyton dealbatum (Thunb.) Hilliard & Burtt*	Compositae	Annual herb	Southern Africa	Australia	N/A
G	*Verbascum thapsus L.*	Scrophulariaceae	Biennial or annual herb	Northern Africa, Europe and Asia	Australia, North America (California)	N/A
G	*Verbascum virgatum Stokes*	Scrophulariaceae	Biennial herb	Europe	Australia	Ornamental
G	*Verbena bonariensis L.*	Verbanaceae	Perennial herb	South America	Southern Africa, Australia	Ornamental
H	*Vinca major L.*	Apocynaceae	Perennial herb	Europe	Australia, North America (California)	Ornamental
Old fields	*Vincetoxicum rossicum (Kleopow) Barbar.*	Apocynaceae	Perennial vine	Europe	North America	N/A
G	*Vulpia bromides (L.) Gray*	Poaceae	Annual herb	Northern Africa and Europe	Australia	Erosion control, revegetating
G	*Vachellia nilotica (L.) P.J.H. Hurter & Mabb. Subsp. Nilotica*	Leguminosae	Deciduous tree	Africa	Australia	Agroforestry, boundary, fuelwood, wood, revegetator, fodder
G	*Watsonia meriana (L.) Mill.*	Iridaceae	Perennial herb	Southern Africa	Australia	Ornamental
G	*Watsonia versfeldii J.W. Mathews & L. Bolus*	Iridaceae	Perennial herb	Southern Africa	Australia	Ornamental
G	*Xanthium spinosum L.*	Compositae	Annual herb	South America	Southern Africa, Australia	N/A
G	*Xanthium strumarium L.*	Compositae	Annual herb	North America and Europe/Asia	Southern Africa, Australia, parts of Europe	N/A
G	*Zantedeschia aethiopica (L.) Spreng.*	Araceae	Perennial herb	Southern Africa	Australia, North America (California)	Ornamental
P	*Ziziphus mauritiana Lam.*	Rhamnaceae	Evergreen shrub, tree	Asia	Australia	Ornamental, fruit, fuelwood, wood

(continued)

Table 7.1 (Continued)

Habitat	Scientific name	Family	Life Form	Native	Continent Invasive	Commercial uses
G	Elaeagnus angustifolia L.	Elaeagnaceae	Deciduous shrub, tree	Europe and Asia	North America, Europe	Honey production, erosion control, ornamental, revegetator, shade/shelter, food, wood
G	Parkinsonia aculeata L.	Leguminosae	Semi-evergreen shrub, tree	North and South America	Africa, Australia, small part of Europe	Erosion control, honey production, ornamental, fuelwood
G	Pinus elliottii Engelm.	Pinaceae	Evergreen tree	North America	South America, Southern Africa, Eastern Australia	Fibre, gum/resin, wood, revegetator, soil erosion control
G	Pinus halepensis Mill	Pinaceae	Evergreen tree	Southern Africa and Northern Europe	South America, Southern Africa, and Western Australia	Erosion control, ornamental, fuelwood, gum/resin, wood

selected, but were introduced for erosion control, as forage species, or as ornamentals. Mack and Lonsdale outlined three time periods in species introductions, the accidental phase, the utilitarian phase, and the aesthetic phase, which is ongoing. The **accidental phase** occurred when species accidentally arrived with human settlement. Native Americans called *Plantago major* "Englishment's foot" because it was found around European settlements in North America. The **utilitarian phase** involved deliberate introduction of species as forage species, for erosion control, and for game species or furbearers. European cattle were introduced widely, furbearers like nutria were introduced into wetland areas, and many grasses and legumes were introduced (Mack and Lonsdale 2001). The bluegrass state of Kentucky in the USA. is a result of the introduction of *Poa pratensis* into the USA, which replaced the savannah and open woodland species in the area (Mack and Lonsdale 2001). Finally, the **aesthetic phase** involved introducing species that are attractive to humans as ornamentals. The European starling was introduced into central park of New York City to include all species mentioned in William Shakespeare's plays.

The implications of these introductions are that the exotic species are not random selections from the source species pool (Alpert 2006, Wilsey and Polley 2006, Wilsey et al. 2015), hence their effects on ecosystems might be greater than if they were random escapes. Pysek et al. (2014) found that species that were widely seeded or cultivated had a larger range in their invaded range than other species. The clover (*Trifolium*) species that were introduced by humans were found to be more problematic that species that escaped on their own in New Zealand (Gravuer et al. 2008).

As a result of human introductions, exotic species have different traits on average than native species that they have replaced or invaded. Exotic species have found to have higher trait values, on average, than native species (Liao et al. 2008), in for example, specific leaf allocation (SLA), higher allocation to aboveground structures and sexual reproduction, higher growth rates, and greater phenotypic plasticity compared to native species. These differences are based on averages, and there are exceptions to the rule. Key traits that affect species interactions, such as growth rate, timing of growth, and reproduction (phenology), and response to variable resource availability may differ between native and exotic species (e.g., Willis et al. 2010, Pan et al. 2010, MacDougal and Turkington 2005). Daehler (2003) reviewed studies that found that exotic species showed a more plastic response to altered resource availability than natives, which suggests that they might respond differently to climate change. A better understanding of if and when there are functional differences between natives and exotics will lead to improvements in global climate change models in situations where the proportion of the landscape that is exotic-dominated is increasing over time (Martin et al. 2014). Pysek et al. (2014) found that range size in the invaded range (North America) was a function of cultivation (planting and selection of genotypes with higher vigor), and range size in their native range, and not a function of commonly measured leaf traits.

Exotic species affect native species in grasslands in many ways. They can reduce native species establishment through **resource pre-emption**, where exotic species capture water or nutrients before native species establish, preventing successful colonization by native species (Yurkonis et al. 2005, Polley et al. 2006). Resource pre-emption can be enhanced when the exotic species arrives or greens-up before the native species (Firn et al. 2010, Dickson et al. 2012, Wilsey et al. 2015). Exotic species are sometimes better competitors than natives because they have been released from their enemies, which frees up resources that can be directed towards competitive ability (**EICA hypothesis**, or Evidence for Increased Competitive Ability, Blossey and Notzold 1995). This partially accounts for the success of Chinese tallow trees (*Sapium sebiferum*) invading coastal grasslands in the southeastern USA. In some cases, such as with members of the *Centaurea* genus (knapweeds), the species produces novel allelochemicals (Callaway and Ridenour 2004). Grasses that are native to the USA, are more strongly affected by these novel chemicals than grasses from the native range of *Centaurea*. Comparing species in their native and invaded range has

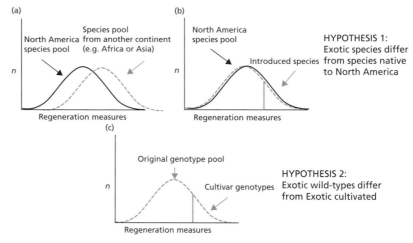

Figure 7.5   Introduced species in North America could have higher regeneration measures because of differences among continents (A). Alternatively, if no differences exist among continents, differences could occur due to non-random choosing of species from the tail of the distribution to introduce (B). Because species were introduced to prevent erosion, for aboveground forage, or as ornamentals, they are predicted to have higher seedling emergence rates, earlier emergence, and higher seedling growth rate, and stronger priority effects as a result. Cultivation is predicted to enhance these measures within species due to human selection in cultivated genotypes compared to wild-types (C). These ideas were tested in Wilsey et al. (2015) with 28 perennial grassland species. There was support for differences among native and exotic species in how they suppress species diversity, but no differences were found between cultivated and non-cultivated genotypes.

helped us to develop a much better understanding of exotic species impacts on grasslands.

7.9.1 Cultivars

Cultivars are cultivated varieties of either native or exotic species. The level of cultivation can vary from merely knowing the location of the seed collections, to selection for enhanced traits (Figures 7.5, 7.6, 7.7). Cultivated varieties exist for many grassland species, and they were widely planted in the past. Cultivars have beneficial properties such as high germination, growth and sometimes late flowering that make them valuable in agricultural settings. However, in restorations and other non-agricultural settings, they can differ genetically in some cases from wildtype genotypes (Gustafson et al. 2004a), and they may differ in their effects on subordinate species in their environment. The **Cultivar Vigor hypothesis** suggests that human selection for increased vigor will lead to increased resource capture and productivity

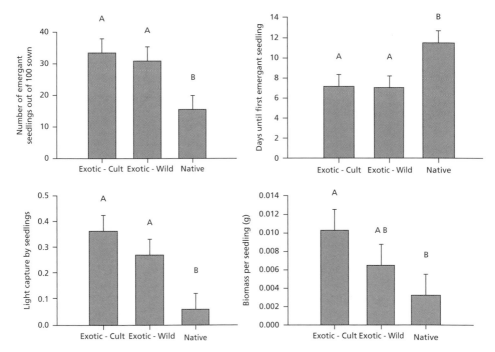

Figure 7.6   Regeneration measures (seedling germination rate, days to first emergence, and light capture, and biomass per seedling, means + SE) of exotic species introduced as cultivars (Exotic-Cult), exotic species from their native range (Exotic-Wild), and species Native to North America (28 species, 14 species from each group). Germination rate (no./100 seeds) was higher in exotic than native species. Days to first emergence was the number of days until the first seedling appeared, and it was shorter in exotic than native species. Light capture and biomass of seedlings was also higher in exotics.

compared to non-cultivars (Wilsey 2010). Alternatively, cultivars are usually collected from seed sources far from the location they will be seeded into, i.e., they are not local. The **Local Adaptation hypothesis** suggests that cultivars will have lower resource capture and productivity when and if local adaptation is important (Wilsey 2010). Reed et al. (2018) found that cultivars of *Pascopyron smithii* had higher growth rates and were better competitors than non-cultivar genotypes. Wilsey (2010) found that germination rate was higher in cultivars than non-cultivar native species in 4 out 5 cases examined. However, cultivars and non-cultivars do not always differ in their aboveground productivity or in how they suppress subordinate forb species (Gibson 2009, Wilsey 2010). Cultivar vs. non-cultivar plantings of tallgrass species had equal NPP, N mineralization rates and soil carbon accrual (Baer et al. 2014). Gustufson et al. (2004b) found higher aboveground biomass production in the cultivated North American grass big bluestem (*Andropogon gerardii*) than non-cultivated genotypes. Wilsey et al. (2015) found that the differences between cultivated genotypes and non-cultivated genotypes is smaller than the native-exotic difference in 28 grassland species studied. The evidence so far appears to suggest that in many cases, the negative effects of being a non-local genotype may cancel out the benefits of having high vigor.

7.9.2 Novel ecosystems

Communities dominated by exotics are considered to be 'novel systems' (e.g., Hobbs et al. 2006) because they contain species from a variety of regions that do not have an evolutionary history of interaction. Novel ecosystems are common in agricultural areas and in suburban/urban areas where human impacts are high, and are especially common in grassland systems (Kulmatiski 2006). Novel ecosystems are predicted to become more common in the future as: 1) exotic species spread into or replace native systems, and 2) species move at different rates due to climate change (Wilsey et al. 2011). Both of these scenarios will produce conditions where species interactions are between evolutionarily naïve species with no natural analog.

Species interactions among evolutionarily naïve species may be fundamentally different from interactions among native species that have a longer history of interacting (Wilsey et al. 2009, Isbell and Wilsey 2011). Wilsey et al. (2009, 2011) found that interacting exotic species had less niche partitioning and complementary resource use than interacting native species. This is probably due to the recent arrival of exotic species. Native species have been interacting for thousands of years (since the last ice age) and have evolved with other species in mixture. Novel grasslands of the Great Plains have higher forage nutrient contents, lower species diversity at the alpha scale, higher beta diversity across latitudinal gradient, and altered productivity levels compared to native sites (Martin et al. 2014, Martin and Wilsey 2015). Novel grasslands also had earlier green-up dates in the central and northern

Figure 7.7   *Hyparrhenia rufa*, a grass native to east Africa (where this picture was taken) that has invaded and spread widely in South America. This species is included in Table 7.1, a list of grassland species that have been documented to have a negative environmental impact. Photo: Brian Wilsey.

Plains sites, and later senescences across the gradient (Wilsey et al. 2017). Mycorrhizal colonization rates can be higher in exotic species than native species, but their effects on plant growth can be less beneficial in exotic species than in native species (Checinska Sielaff et al. 2018). Native and exotic species are not ecologically interchangeable in grassland systems, and novel ecosystems are probably here to stay.

Among the most noxious grassland invaders is the red imported fire ant (*Solenopsis invicta*). This ant was introduced into the USA in the late 1930s, and has recently invaded into Australia, New Zealand, China, India, Malaysia, and the Philippines. Fire ant abundance is favored by disturbance, and mounds can be especially plentiful in heavily grazed and formerly plowed grasslands. Fire ants have a swarming behavior when animals contact the mounds, and they have been known to kill fawns of white-tailed deer and rabbits (Allen et al. 2004). Ant communities are highly altered by fire ants and are less species diverse in infested areas. Gotelli and Arnett (2000) found that fire ants have altered the latitudinal gradient in species richness of ants, with southern areas with fire ants having fewer species than northern areas without fire ants. Areas with fire ants had an ant community that went from structured to random (Gotelli and Arnett 2000). The disruption by fire ants on the latitudinal species richness gradient for ants was similar to what Martin et al. (2014) found for the latitudinal gradient for C_3 plants (Figure 4.4). Finally, predators of native ants, like the Texas horned lizard, are negatively affected by the decrease in abundance of native ants caused by fire ants (Allen et al. 2004). Fire ants are currently being displaced by a new exotic ant from South America, which will alter the fire ant effects in the future.

8 Conservation and Restoration of Grasslands

Much of the world's cropland is on areas that were formerly grassland, and many large cities were placed in grassland areas. Tallgrass systems have been replaced by corn-soybean rotations, and mixed and short grasslands have been replaced by wheat. This is due to the high fertility of soils in former grassland sites. Grasslands do not store C in long-lived aboveground biomass as trees do, so much of the carbon ends up being stored in the soil as organic matter (SOM). SOM is negatively charged and can hold cations in cation exchange sites, is a source of slowly released nutrients, and can improve moisture conditions by adsorbing water. The high fertility soils make grasslands excellent for farming. Grasslands are correctly referred to as the "bread baskets" of the world due to their ability to feed humans. Annual crop production can lead to higher rates of soil erosion, runoff, leaching of nutrients into waterways, and loss of soil organic matter. Because grasslands have been so heavily impacted by humans, grassland conservation and restoration to address these environmental issues have been active areas of research. Grassland conservation usually entails the development of grazing plans that mimic natural processes in terms of grazing intensity and timing. Restoration usually entails replacing domestic livestock and plants with native species.

8.1 Conservation of grasslands

Conservation of grasslands usually involves adjusting management plans to consider grazer stocking rates and composition of grazing species and seed mixes. Livestock grazing is more compatible with the maintenance of ecosystem services in intact native grasslands than is crop production if livestock are properly managed. For example, Burke et al. (1989) found that soil C contents were much higher in areas managed as rangelands than crop fields when compared at similar rainfall levels and annual temperatures. Lower stocking rates, and movement of livestock over time to mimic natural grazing

The Biology of Grasslands. Brian J. Wilsey. Published 2018 by Oxford University Press. © Brian J. Wilsey 2018.
DOI 10.1093/oso/9780198744511.001.0001

patterns, can be highly compatible with the maintenance of biodiversity in native dominated grasslands. This can be quantified with grazing intensity, or the net primary productivity that is consumed/NPP. **Grazing intensity** has two components, the **frequency** of grazing and the **proportion of tissues that are removed** during a grazing event. Grazing is predicted to have the smallest impacts on native systems when grazing intensity is low to moderate, when the system evolved under high grazing pressure (Milchunas and Lauenroth 1993), and when grasslands have time to recover between grazing events (low to moderate grazing frequency, Oesterheld and McNaughton 1991). However, grazing that is too intense or incorrectly applied can lead to degradation. Degraded systems have a high proportion of annual or unpalatable weedy species, a high proportion of invader species, or both.

Domestic and wild grazing systems can differ significantly, and conservation plans usually promote livestock management that is similar to that of wild grazing systems that have persisted for thousands of years. Livestock systems usually have an order of magnitude higher grazing pressure than native systems due to animal husbandry practices (Oesterheld et al. 1992). Shifting from a native collection of grazing mammals to domestic animal counterparts led to altered grasslands and lower C storage in India (Bagchi and Ritchie 2010). High stocking rates can shift perennial native-dominated grassland to exotic- or annual-dominated grassland. Differences are smaller when comparable stocking rates are used, but subtle differences still exist. For example, cattle eat slightly more forbs and spend more time near water bodies than native bison species (Towne et al. 2005, Allred et al. 2011).

A second objective of grassland conservation programs is to find and protect the remaining areas that have not been plowed or converted to simplified crop systems. Because of the massive conversion of many native-dominated grasslands to agriculture, unplowed remnants exist as small scattered areas embedded in an agricultural or urban/suburban matrix. Grassland conservation usually entails identifying and protecting the remaining areas and developing management plans to assist with their long-term maintenance. Grasslands are typically in areas that are highly suitable for agriculture. Protected areas like national parks are found in areas with interesting geological history and are typically in areas that are not suitable for agriculture. Because of this, grasslands traditionally were not set aside for protection as often as forests and deserts (Samson and Knopf 1994, Veldman et al. 2015). Non-government organizations like the Nature Conservancy and government agencies have rectified this in recent years by trying to protect some of the remaining high-quality remnant grasslands (Ricketts et al. 1999).

Locating remnants is harder than it sounds. Some perennial-dominated areas that look like they have not been plowed sometimes are full of exotic species. In this case, it is hard to differentiate a remnant from a perennial community that assembled after abandonment of agriculture. The concept of the floristic quality index has been developed to help in identifying native grassland remnants (Box 8.1).

Box 8.1 Floristic quality index

FLORISTIC QUALITY INDEX: The floristic quality index is calculated as: $C \times (\sqrt{N})$, where C is the conservation value of a species, and N is the number of plant species at the site. A "conservative" species with a high C value is solely restricted to remnant habitat, whereas species with a low C value are found in degraded environments only. The presence of species that are only found in remnants is a good indication that the site has not been plowed or overseeded. However, the C value of each species has to be estimated with field data or from expert opinion, and the C values are not known for all species. The assumption behind this approach is that the C value is an indicator of true remnant prairie species. For example, many prairie forbs that are only found in remnants have a value of 10, and species that are found virtually everywhere, whether remnant or not, receive a value of 1. Richness (N) is included in the formula because the quality of a remnant is usually associated with having high species richness. With this approach, many of the remaining remnants in heavily degraded areas have been identified and protected in the Midwestern U.S.A. The floristic quality index (FQI) is also used to assess the quality of restorations.

Formal grassland conservation programs vary by country. In the USA, the Conservation Reserve Program (CRP), the objectives are to manage for the following environmental benefits through the use of perennial plantings:

1. Wildlife habitat cover
2. Water quality benefits from reduced erosion
3. On-farm benefits of reduced erosion
4. Enduring benefits
5. Air quality benefits
6. Low cost

A total of 42 seed mix types have been developed since the program was founded in the 1985 farm bill. Most of the original mixes were introduced grasses only (e.g., *Bromus inermis*), native grasses only (e.g., *Panicum virgatum*), or simple mixes of grasses and the legumes alfalfa *Medicago sativa* and birds foot trefoil *Lotus corniculatus*. The original objective was to reduce erosion and replenish soil organic C. Recently, other objectives such as increasing pollinator and native species habitats have been incorporated into projects. Over time, seed mixes have been more targeted to the local vegetation types and have become more complex and similar to full restorations. For example, the conservation planting CP42 is a pollinator habitat seed mix that must include at least nine species that will flower in the early, mid and later parts of the growing season. As of 2016, 23.5 million acres (9.5 million ha) of land was enrolled in the Conservation Reserve Program. In Europe, low-intensity land use such as no fertilization and cutting only once per year are used in some conservation plantings to provide habitat for native species and to maintain biodiversity in rural landscapes.

Biofuel plantings typically use perennial grasses such as switchgrass, *Panicum virgatum*, *Miscanthus* spp., and grassland mixtures to produce ethanol fuel. Perennial plantings such as these have the potential to produce multiple environmental benefits (Schulte et al. 2017). If they use diverse seed mixes, they can achieve the same environmental benefits as restorations (Jarchow and Liebman 2012).

8.2 Grassland restoration

In many situations there is an interest in restoring former crop lands or heavily grazed invaded pastures back into native grasslands. This is common in parkland sites, in recreational areas near urban and suburban areas, in hilly or marginal environments on highly erodible soils, in buffers along waterways, and similar situations. Restoration is defined by the Society of Ecological Restoration (SER) as the 'improvement of an area that has been degraded by humans ....' Successfully restored sites are considered by the SER to have a high proportion of indigenous (native) species, to have the major functional groups present, to be functioning properly, and to be integrated into the surrounding landscape. These objectives are more difficult to achieve than they sound, especially the goal of successfully integrated restorations into the surrounding landscape for perpetuity. Grasslands in subhumid and humid areas require human inputs (burning or grazing) to persist as a tree-free system.

8.2.1 Restoration as an acid test for theory

Attempting to restore an area using ecological theory sometimes reveals previously unrealized problems with the theory. Ecological restoration has been described as an Acid Test for ecology by Tony Bradshaw (Bradshaw 1987). If we fully understand how grassland systems function and assemble after disturbance, then it should be easy to restore them after they have been degraded or destroyed, right? During the process of restoration, we often learn that we do not fully understand how systems function and assemble. Many failed restorations led to research into underappreciated fields, such as soil processes and plant-soil feedbacks (Falk et al. 2006).

8.2.2 Problems with changing climate and species pools

Traditionally, restoration plans have followed the classical view of plant species succession (Kuchler 1964). This approach assumes that climate conditions will favor the eventual establishment of the potential vegetation type if given sufficient time (Clements 1916, Figure 8.1). For example, using this framework, managers typically allow the area to recover after disturbance along expected successional trajectories if the disturbance was light, or in cases where the disturbance was moderate, they might add seeds of native species to speed up succession. Having reference sites or past conditions is helpful in this regard (White and Walker 1997, Polley et al. 2005, Martin et al. 2005). However, new realities such as an altered and changing climate and the presence of exotic (non-native or 'invasive') species complicate this approach (Figure 8.1).

Restoration targets based on potential vegetation become problematic in a changed climate. Restoration plantings will have difficulty matching expectations from potential vegetation maps (Kuchler 1964), partially because the climatic conditions of the past are now gone (Polley et al. 2014). It will be impossible to restore the atmospheric CO_2 concentration, which is now approximately 400 ppm, to preindustrial levels of 270 ppm. Increased CO_2 concentrations have far-reaching effects (Chapter 7). Temperatures have increased by 1°C since pre-industrial times as CO_2 has increased and are expected to continue to increase by 2–3°C during the twenty-first century (IPPC 2014). Rainfall is becoming more variable everywhere, and will be lower in many areas (e.g., low-latitude sites), and higher in others (e.g., high-latitude sites). The changing climate makes restoration of grassland systems to pre-settlement conditions very challenging if not impossible, as potential vegetation of the past may be mismatched to future climates.

Furthermore, the species pools available for colonization of bare ground after disturbance have changed from purely native species to mixtures of native and non-native species (Figure 8.1). This will limit **passive restoration** in areas that have a high abundance of non-native species. In many cases, quickly establishing non-native species have been seeded in areas that were heavily disturbed to prevent erosion in the past, and in some cases, these species have persisted for long periods of time and have caused declines in native species propagules (Bakker and Berendse 2002, Weber 2017).

Thus, the vegetation that develops following disturbance in areas far from native propagule sources may be very different from the potential vegetation

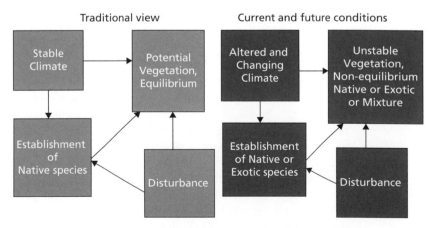

Figure 8.1   Relationships between climate and vegetation are hypothesized to change over time with changes in the stability of climate (shift from a stable climate to a changing climate) and biota (all native species to native or exotic species or mixtures of both). This new reality has to be better understood in order to develop future restoration plans.

type. The native seed bank will be depleted in areas where native grasslands exist as small patches in a larger matrix of non-native species and in areas that have undergone transformation for long time periods (Bakker and Berendse 2002). As a result, the species pool available for establishment often contains a mixture of native and non-native species in varying proportions or a new combination of native species that we have not seen in the past (Figure 8.1). Climate change will lead to additional reorganization of communities as species move at different rates across landscapes.

In some cases, the communities that assemble in modern environments are different enough from past communities to be termed 'no-analogue' or 'novel' ecosystems (Hobbs et al. 2006). They are novel in that they differ from what we have seen in the past and are different from what is expected based on the potential vegetation concept. We currently have a poor understanding of these new community types and developing a greater understanding will be extremely important in informing current and future management. Wilsey et al. (2009, 2011, 2017) and Martin et al. (2014, 2015) have found that novel grasslands of the central U.S.A. have lower plant diversity and altered species interactions compared to the native grasslands that they replaced. Thus, restoration plans that take into account climate change and non-native species must be geared to tipping the balance from exotic towards native dominance rather than focused on a complete restoration to a pre-settlement state.

There have been surprisingly few studies on what tips the balance towards native species in restorations. Soil N reduction led to greater establishment of native species and a higher Native:Exotic ratio in an early study (Blumenthal et al. 2003). Suding et al. (2005) found that exotic plant species are favored by N additions. Van der Heijden et al. (2008) found that soil feedbacks can explain native species dominance.

8.2.3 Can grasslands be fully restored to reference levels?

This question is an essential question in restoration ecology. This issue comes up commonly with managers, and forms the foundation of mitigation efforts, where ecosystem destruction at one location is replaced by a restoration planting at another place. The author was at a meeting once where a loss of a remnant prairie for a road expansion in the U.S. was being discussed. The manager explained that the roadside would be seeded with a prairie mix after the project was completed. Is the plant community that develops after seeding equivalent to the remnant prairie that was found there originally? If grasslands can be fully restored to conditions that we find in non-degraded reference sites, then it will less important to protect the remaining non-degraded intact remnant sites. If the answer is no, then it remains imperative to fully protect remnant areas because they are irreplaceable.

The idea that restorations will not be equivalent to remnants has been termed the **'Humpty Dumpty' hypothesis** (Pimm 1991)—once a system is lost, it cannot be put back together again. The alternative hypothesis suggests that restorations will achieve remnant levels if given sufficient time and attention. This alternative hypothesis is a major assumption of most restoration efforts: values of biodiversity, species composition, and ecosystem functioning that exist in remnant areas can be restored if given a sufficient amount of time (Figures 8.2, 8.3). This hypothesis has been tested with three approaches: 1) a comparison of restorations and remnants at a given time, 2) with chronosequence studies, which compare different aged plantings at a given time, or rarely, 3) with time course studies that follow restorations over long time frames (Table 8.1). The first two approaches are much more

Figure 8.2   A range of possible outcomes are possible from restorations, from more successful, diverse plantings that are similar to remnant areas (top) to less successful plantings that have low diversity and poor establishment from the seed mix (bottom, dominated by the non-seeded exotic species *Bromus inermis*). The restoration on the bottom was seeded with 52 plant species, but is dominated by only one. Photos: Brian Wilsey.

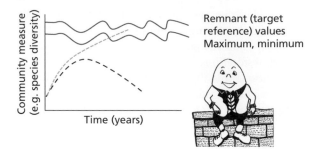

Figure 8.3   Hypothesized time course for restorations. The top dashed line supports the idea that restorations will achieve remnant/reference levels given sufficient time. The humped line supports the 'Humpty Dumpty' hypothesis that restorations will not achieve remnant levels. Species diversity will peak after a few years, and then will decline in later years, never achieving the levels of remnants. In this case, the degraded state is an alternate state to the reference state. Figure modified from Wilsey and Martin (2005).

Table 8.1 Studies that have measured species diversity and ecosystem measures over time with a chronosequence study (prediction is that it should increase over time to reference levels) or by comparing restorations to remnants (prediction is that it will equal reference sites). A non-significant difference between remnant and reference levels is denoted by a =. Note that only one paper is included per study.

Study	Details	Age (years)	Results: Remnant vs. Reference
Chronosequence studies:			
Brye Riley (2009)	Soils only	3, 4, 5, 26	Nutrients declined
			Soil OM, C: increased
Sluis (2002)	Former crop	1–15	Plant species richness lower in restorations than remnants
	Multiple plantings		Species richness declined over time
Cammill et al. (2004)	Former crop	1–8	ANPP—no change
	No remnants		C, N—no change
			Plant species richness highest at intermediate
Baer et al. (2002)	CRP	1–12	Forbs declined dramatically over time
			ANPP, C, N—no change
			ANPP—restoration > remnant
			BD decreased
Grman et al. (2013)	27 restorations	3–7	Plant species richness declined over time
			Sown species richness declined non-significantly over time
Hansen and Gibson (2014)	Literature Reference	2–19	Mean C: restoration = remnant
	Crops, remnants		FQI—4 of 19 achieved reference levels
			ANPP—13 of 19 achieved reference levels
			N, C, Bulk density—few achieved reference
			Simpson's diversity—declined over time

Study	Details	Age (years)	Results: Remnant vs. Reference
Comparison to remnant studies:			
Kindscher Tieszen (1998)	2 restorations	5, 35	Plant species richness: remnant > restoration
	2 remnants		Plant species diversity (H'): remnant > restoration
			Plant species diversity (H'): remnant = restoration
			Soil C: remnant > restoration
Brye et al. (2002)	1 restoration	19–23	Soil C, N: restoration < remnant
	1 remnant		ANPP no change over time
			Mean C: restoration = remnant
			Plant species richness: restoration = remnant
Polley et al. (2005)	3 restorations	9, 10, 20	Plant species richness: restorations < remnants
	3 remnants		Plant species diversity: restorations < remnants
			Exotic species biomass: restorations = remnants
Martin et al. (2005)	8 restoration plantings	7–9	Plant species richness: restoration < remnants
	3 remnants		Plant species diversity: restoration < remnants
			ANPP: restoration > remnants
			Exotic species biomass: restorations = remnants
McLachlan Knispel (2005)	9 restorations	3–15	Plant diversity: restorations < remnants
	3 remnants		Exotic plants: restorations > remnants
			Plant diversity declined over time
			Soil pH: restorations = remnants
			Soil P: restorations > remnants
			Soil N: restorations = remnants
Kucharik et al. (2006)	1 restoration	65	Plant species richness: restoration = remnant (15.8 vs. 14.1 /m^2)
	1 remnant		Soil C 37% higher in remnant than restoration
			ANPP: restorations = remnants
Matamal et al. (2008)			stocks 95% of remnant level
Carter and Blair (2012)	Former	4, 6, 7, 9, 10	Plant species diversity declined over time
	Pastures,		Plant species richness: restorations = remnants
	Mowing		Mean C: restorations < remnants
	Frequent		
	Fire		

common than the third. Chronosequence studies are harder to interpret than some realize, because the way sites are restored sometimes changes over time. Plantings have to be done in a similar manner over time for it to be useful.

8.3 Ecological assembly rules

Using ecological theory to guide restoration ecology offers a way forward. If general theory is being tested with restorations, it will be more generally useful to the greater ecological community.

Many in the restoration community have argued that to understand how to restore a grassland system, one must develop and test community assembly rules (Weiher and Keddy 1999, Temperton and Hobbs 2004). Community assembly may follow rules, and if these rules are uncovered, then we may be able to accurately predict final species composition after assembly (Figures 8.4, 8.5). Finding the traits behind assembly rules has been described as the

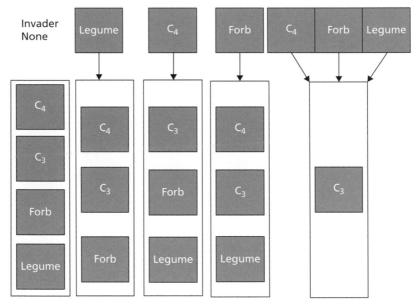

Figure 8.4   Assembly rule 1: the limiting similarity hypothesis and the functional group (or guild) assembly rule predict that the presence of a functionally similar species (in the same guild or functional group) will restrict invasion by other members of that group. Boxes with arrows denote invaders or colonizers. Functional groups in this case are C_4 grasses, C_3 grasses, forbs, and leguminous forbs, but this applies to any functional groupings or animal guilds. It can also be based on quantitative traits. The loss of that group will lead to greater invasion rates by members of that group. The low diversity community on the right has only C_3 grass species and is invaded by all other functional groups.

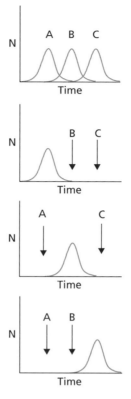

Figure 8.5   Assembly rule 2: the temporal assembly rule predicts that invasion or colonization will be low for species when species are present that are actively growing (denoted by species A, B, and C) at the time of disturbance. Arrows denote invaders or colonizers.

Holy Grail of community ecology by Weiher and Keddy (1999). In grasslands, evidence supports the **limiting similarity hypothesis**, which predicts that a member of a specific functional group (usually C_4 grasses, C_3 grasses, legumes, forbs) will have greater difficulty establishing into an area dominated by another member of that specific functional group. For example, a patch dominated by C_3 grasses is predicted to be more likely to be invaded by C_4 grasses, legumes or forbs than by another C_3 grass. This idea was largely supported by Fargione et al. (2003) with propagule additions into experimental plots in Minnesota. In general, there was a negative correlation between establishment by a member of a functional group, and the biomass of that particular functional group in the plot. A related idea was proposed by J. Bastow Wilson, who suggested that the proportions of these functional groups should settle on specific values given enough time (Wilson 1999). Mowed grasslands from different parts of the world tend to have similar proportions of each functional group (Weiher and Keddy 1999). Another assembly rule is the **temporal assembly rule**. The timing of a disturbance within a year could lead to a separate species composition (Keever 1950, Questad and Foster 2008, Martin and Wilsey 2012). For example, a local disturbance in the late summer or fall should favor short-day plants with C_3 photosynthesis, whereas a local disturbance in later winter or spring should favor long-day plants with C_4 photosynthesis. Thus, depending on the timing of the local disturbance, we can predict the species composition of the assembling community (Martin and Wilsey 2012). Keever (1950) found in NC old fields that plowing in summer led to horseweed and aster dominance, whereas plowing in the late fall led to ragweed dominance in the following year. The presence of these species might in turn, influence the later establishing species. Howe (1994, 2011) found that the season of burning was influential in explaining the species composition of prairie plantings in the central U.S. Spring burning led to a reduction in early flowering species and an increase in C_4 grasses. Summer burning led to the opposite effect.

Once communities form, major models differ in predicting how stable they will be and what will maintain them over time. The state-and-transition model predicts that changes in species composition due to major disturbances (e.g., heavy grazing) can result in alternate states of vegetation, with feedback occurring within states to reduce the likelihood of transitions (Westoby et al. 1989, Briske et al. 2005). Under this model, human degradation or heavy grazing causes shifts in species that are tolerant of heavy disturbances. Transitions can occur from one state to another once thresholds are passed. These new states may be fairly persistent over time. If they are not persistent, they may occur as transient states (Fukami and Nakajima 2011). The second group of models predicts that variation in species traits leads to predictable and often linear changes in species composition. These models make predictions about species diversity as well as composition. The **competition-colonization model** predicts that a trade-off occurs between competition and colonization ability in individual species (Pacala and Rees 1998). With no disturbance, the best competitor will become dominant given sufficient time. Thus, occasional disturbance enables colonizers to coexist with competitors. Other models are based on the level of symmetry in size or reproduction between species, and predict that species coexistence will be least likely when differences in fitness occur between species, i.e., the **equalizing effect** of Chesson (2000). Furthermore, under this model, disturbance increases diversity if it has a greater impact on abundant species and decreases it if it has a greater effect on rare species. Differences in niche requirements among species can also lead to a stabilizing force that can counter equalizing forces (Chesson 2000). Related pure niche models predict that species are adapted strongly to resource availability, so species that are adapted to disturbed conditions will be favored by conditions shortly after disturbance (Pacala and Rees 1998). With time since disturbance, the species adapted to low resource availability increase in abundance. Finally, the **neutral model** predicts that assembly will be largely

random in nature and will depend on ecological drift or random fluctuations of species based on the identity of individuals near a disturbed
patch (Hubbell 2005).

8.4 Priority effects and alternate states

Succession theory predicts that the sequence of species arrival order during
recovery from a disturbance should have little effect on the eventual composition of the community; that is, the community will eventually reach a
climax. The size differences of competing species should not matter, and early
models emphasized facilitation only (Clements 1916). Once the late successional species arrives, it will colonize and become locally dominant. Species
with traits that are helpful in colonization, but that are not good competitors
will be restricted to areas of recent disturbance. They will persist as fugitive
species by re-emerging in areas with a recent disturbance, perhaps disturbed
by digging animals (Platt 1975). This leads to a mosaic of successional stages,
all being regulated by the appearance of new disturbed patches.

However, beginning in the 1960's ecologists realized that arrival order did
affect the long-term outcome of species interactions (Harper 1961). Early-
arriving species can facilitate, suppress or have neutral effects on later-arriving
species (Connell and Slatyer 1977). Early-arriving species can filter out
species from the colonizing species pool and can alter the eventual species
composition (Young et al. 2005, Fukami 2015). The first grassland studies on
this were done by John Harper and his lab (e.g., Harper 1961). They found
that relative abundances between species could be drastically changed
depending on which species seeds arrived first. When seeded at the same
time, a *Bromus* species (*rigidus*) made up 75 percent of total biomass over a
second species (*B. madritensis*). This value was reduced to 10 percent when
B. madritensis was seeded 21 days before *B. rigidus*. Körner et al. (2008)
found strong priority effects in grassland communities in a greenhouse setting, and that the priority effects could not be overturned by clipping. Young
et al. (2017) reviewed priority effect studies and found that the majority of
studies were short term, but that most found strong support for priority
effects. Longer-term field studies were less common in the literature, but
priority effects either persisted over many years (Martin and Wilsey 2012,
2014) or strengthened in time. Priority effects are also likely to be context
dependent, with their effects likely to be stronger in fertile environments.

One strong predictor of priority effects is the native-exotic status of the species that established first (Wainwright et al. 2012, Dickson et al. 2012, Wilsey
et al. 2015). Dickson et al. (2012) conducted a greenhouse experiment with
three native-exotic pairs and found that three exotic species formed near

monocultures when seeded 21 days before other species. The three native species were colonized and developed into diverse communities when seeded 21 days before other species. Wilsey et al. (2015) found a similar result when using 14 native-exotic species pairs (28 species total). Exotic species suppressed establishment from a 39-species seed mix when seeded 21 days before, and interestingly, this suppression was found both with exotic genotypes that have been cultivated and 'wild' genotypes. Exotic species had higher germination rates than natives and emerged more quickly from the soil than native species (Figure 7.6). Goodale and Wilsey (2018) found that exotic species priority effects were consistent across rainfall variability treatments, suggesting that the exotics were phenotypically plastic. These results suggest that natives might be favored in restorations in subhumid grasslands if they are established before exotic species invasions occur.

8.5 Linking metacommunities—seed limitation and shifting limitations

Key questions in restoration ecology are whether propagule additions are necessary or alternately, if passive restoration (allowing succession to occur) will lead to the diversity of remnant sites given enough time. Passive restoration might lead to the full diversity of a remnant site if primary productivity is low, and the site to be restored is near and connected to a remnant (Prach and Hobbs 2008). Restoration in this case may entail the removal of the stressors to the system, adding intrinsic disturbances such as prescribed fire, or ending the force that is degrading the site, and then protecting the area to give it enough time to recover. In large intact areas, restoration typically consists of removing the degradation, and initiating a fire regime that matches the typical fire return interval of the area (Collins et al. 1998).

Originally, native grasslands were extensive, and patches were connected (Figure 8.6). Now, croplands, cities, woodlands, or heavily grazed pasture can be present around native grassland sites, and native sites are commonly isolated and disconnected from other sites. The shrinking of grassland size due to land conversion, and associated changes that have occurred such as altered fire regimes have led to species loss in many remnant sites (Leach and Givnish 1996). Small size has led to inbreeding depression in some forb species (Kittelson et al. 2015). Leach and Givnish (1996) revisited many Wisconsin prairies that were originally sampled by John Curtis and his students, and found that species richness was reduced. Legumes, short-statured species, and originally rare species were now missing in the remaining remnants. Other revisitation studies have found that rare species are more likely to go locally extinct, although in many sites (Wilsey et al. 2005), the level of local extinction is lower than what would be predicted by Island Biogeography theory.

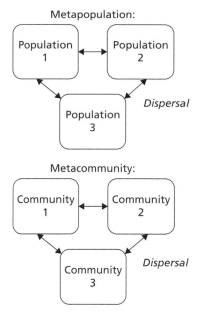

Figure 8.6   Metacommunities contain a network of local communities, much as metapopulations are made up of populations. Dispersal links each community (denoted by arrows), and local extinction can be reversed with enough dispersal. High dispersal can reduce differences among local communities, or beta diversity (Loreau 2000). However, it is important to remember that dispersal rates are not always high among plant communities, and seed additions sometimes have no long-term effect on diversity without disturbance (Foster et al. 2004, Wilsey and Martin 2015).

The species diversity of these isolated sites, as well as new grassland sites that assemble on former cropfields, could be limited by lack of **seeds**, **microsites** for establishing seedlings, or a combination of both. Microsites are openings in the canopy that facilitate seedling emergence and establishment. The **shifting limitations hypothesis** of Foster predicts a shift from seed-limited to microsite limitation as primary productivity increases, and several studies have supported this (Foster et al. 2004, Figure 8.7). Productive grasslands will be hard to restore with pure seed additions in this case. Seed addition studies have found that seeds of native species are often limiting due to a lack of connectance to remnant areas (Foster et al. 2004). Adding seeds experimentally often leads to a short-term increase in species richness, suggesting simple seed limitation. Species richness can be increased in the short term by a few emergent seedlings. Seedlings may not recruit into reproductive populations, however, as is necessary to increase species diversity in the long term. Studies that have followed the seed additions over time have found that microsite limitation becomes more important as time progresses (Wilsey and Polley 2003). Wilsey and Martin (2015) found that grazing by bison became more important over a four-year study. Initially, local diversity

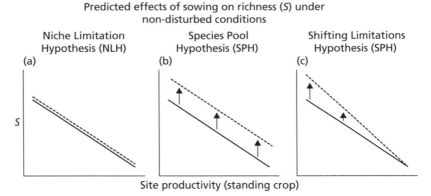

Predicted effects of sowing on richness (S) under
non-disturbed conditions

Niche Limitation
Hypothesis (NLH)

Species Pool
Hypothesis (SPH)

Shifting Limitations
Hypothesis (SPH)

(a)

(b)

(c)

S

Site productivity (standing crop)

Figure 8.7   How the number of species (S) that establish into field sites can be affected by standing crop biomass. The niche limitation hypothesis predicts that few species establish with seed additions (a). The species pool hypothesis predicts that species establishment will be high and independent from site biomass (b). The shifting limitations hypothesis predicts high establishment only in areas with low biomass. Seed additions must be followed for a sufficient amount of time to determine if establishment was successful. From Foster et al. (2004) J. Ecology, used with permission.

appeared to be primarily seed limited, but by the end of the experiment, seed additions only increased species diversity in areas that had moderate grazing by bison. Bison create microsites through their grazing and trampling activities. Consistent with this, seed addition studies commonly find much more recruitment with a disturbance treatment than without (Foster et al. 2004). Further work is needed in this area, but the work so far suggests that seed additions, especially with a disturbance that knocks back the dominant species, will lead to some recruitment of rare species and an increase in species diversity (Williams et al. 2007), at least in low-productivity sites. In highly productive sites, especially those that are heavily dominated by exotic species (Suding et al. 2004, Wilsey and Polley 2003, Wilsey et al. 2011), seed additions alone may not increase diversity without treating the exotic species and the source of high productivity.

8.6 Woody plant encroachment

A major conservation issue worldwide in grassland systems is woody plant encroachment. Woody plant encroachment is the 'increase in density, cover and biomass of indigenous woody or shrubby plants in various grasslands' (van Auken 2000). The processes behind this increase are multifaceted and debated by researchers, but usually altered fire regimes are involved, often due to livestock grazing that lowers fuel mass. Climate change can interact with these other factors, especially in the case where elevated CO_2 favors C_3 trees and shrubs over C_4 grasses in C_4 dominated grasslands (Polley et al. 1997, Bond and Midgley 2000, Morgan et al. 2007).

Future changes in climate may favor increased woody plant encroachment. Polley et al. (1997) found that elevated CO_2 in the atmosphere favored woody *Prosopis* species over C_4 prairie grasses, primarily through its indirect effects. Under the CO_2 concentrations expected in the future, stomata were closed more often, which led to lower transpiration, wetter soils, and more moisture at deeper depths, which favors the woody plants. Morgan et al. (2007) found that a subshrub increased over time in a multi-year global change experiment. Kulmatiski and Beard (2013) found that more intense rainfall in Kruger National Park, South Africa resulted in greater soil moisture in deeper soils that favored woody plants over grasses. This niche partitioning of soil depth usually leads to coexistence in savannas, but this may be dependent on having moderate rainfall intensities. If rainfall becomes more intense in the future, it may lead to greater woody plant growth (Kulmatiski and Beard 2013).

Fires are now extinguished by humans in many parts of the world, and the lack of fire has led to increases in woody plants world-wide. Van Auken (2000) points out that most woody plants that are encroaching into grasslands are native and have contracted and expanded on a geologic time frame. Woody plants were limited to east facing hillsides near rivers when fire was infrequent during presettlement times. Increasing CO_2 concentrations may limit the effectiveness of fire in retarding woody encroachment by speeding tree and shrub growth, allowing woody plants to escape fires by achieving a taller stature more quickly than in the past (Bond and Midgley 2000).

There are large impacts of woody plant encroachment in grasslands (Table 8.2). Species diversity is low beneath woody plants from the genus *Juniperus*. Knapp et al. (2008) found that species richness per 10 m^2 declined from around 30 to below 5 after invasion by eastern red cedar (*Juniperus*) in Kansas, USA. An analysis of 29 studies of 13 different grassland/savanna systems found that plant species richness declined by an average of 45 percent following woody encroachment, and that effects were larger in areas with higher precipitation. Woody plant encroachment in the Brazilia Cerrado due to fire suppression led to 1.2 Mg ha-1 year-1 in C in tree stocks, but to a 27 percent and 35 percent reduction in plant and ant species richness (Abreu et al. 2017). The authors suggested that tree encroachment might lead to a management trade-off between C sequestration and biodiversity.

It is unknown how long encroaching woody plant species can remain before depleting the seed and bud bank of pre-existing species below canopies. Woody plant invasion may represent an alternate state after which return to the native grassland state is impossible. Price and Morgan (2008) found a reduction in species richness of 15 percent in southern Australia with encroachment. Importantly, soil seed banks were depleted in species richness and abundance of grassland species, which suggests that the encroachment will have long-term impacts (Price and Morgan 2008).

Woody propagule pressure is important in the ability of a given fire to suppress woody plant establishment. In very open systems, the distance that a woody plant propagule must travel makes woody plant establishment less likely, even with fire suppression. In many grassland systems, increased human populations sizes and the planting of trees in cities and around homesteads have greatly increased the propagule pressure to the remaining fragmented grassland patches. As a result, the frequency of fires necessary to keep an area woody plant free is much higher than it would have been in the past.

Woody plant encroachment is having important effects on C and water cycling. Biomass is higher in aboveground trunks when woody plants are present, and C is stored in woody plant parts. Woody plants have a greater proportion of C allocation in aboveground parts than belowground parts. Thus, there is a shift in C allocation from belowground to aboveground after trees establish in grasslands. Furthermore, if the tree is evergreen as in *Juniper* spp., they can alter the water cycle as well by transpiring year around instead of during the growing season.

Animal communities are also affected by woody plant encroachment. Grassland bird species are less abundant in areas with *Juniperus virginiana* encroachment, and the community shifts to a more woodland type of community (Coppedge et al. 2004). Small mammals are especially affected, with an increase in *Juniperus* cover from 0 to 30 percent leading to a shift from a species-rich prairie rodent community to a depauparate community dominated by white-footed mice (*Peromyscus leucopus*) (Horncastle et al. 2005), see Table 8.2.

Table 8.2 Selected ecological consequences of the shift from a C_4 grass-dominated to a C_3 shrub-dominated ecosystem (from Briggs et al. 2005, used with permission).

Measure	Shrub island	Grassland
ANPP (g/m²/year)	1284 ± 131	356 ± 28
Standing crop C (g/m²)	3267 ± 465	241 ± 33
Standing crop N (g/m²)	37.8 ± 5.3	6.1 ± 0.8
Surface soil organic C	4740 ± 76	4721 ± 213
Inorganic Soil N (µg per g)	1.81 ± 0.16	9.02 ± 0.65
Annual CO_2 flux	4.78	5.84
Soil CO_2 flux (µmol/m²/s)	5.0–6.2	6.2–8.5
Percent N in leaves	2.84 + 0.06	1.57 + 0.04
Species richness (no. per m²)	11.2 ± 0.9	20.1 ± 1.0

Restoration of these areas involves removing the woody plants with machinery or by hand and burning the trees. Fire usually is reintroduced into the system after trees are removed. Less work has been done on reintroducing browsing and mixed feeder species to restorations (e.g., elk, *Cervus canadensis*), but browsers may be important in limiting woody growth or abundance so that fires are more effective in reducing woody encroachment. Recovery of native grassland species is possible if the trees are removed within a few years of establishment. Beyond a few years post-establishment, seed and bud banks will be depleted, but it is poorly known how long that takes. Long-established tree populations may lead to an alternate state in which grass cover is low and fire behavior is altered so as to have limited impact on trees.

8.7 Savanna restoration

Savannas are grasslands with scattered trees (Figure 8.8), and they grade into woodlands, which are usually not considered a grassland type. They are stable and are not merely grasslands that have suffered tree encroachment. Savannahs are tremendously diverse grassland-type systems because they have an open grassland fauna and flora as well as a separate flora and fauna found under trees (Figure 8.8). Savannas are widespread in Africa, with more than half the continent covered with savannah. Savannah is also common in the Cerrado region of Brazil, in Australia, in parts of South America, Asia, and in central and the southwestern USA. Savannahs are difficult to define precisely, but most definitions describe savannahs as having a grassy

Figure 8.8   Savanna in east Africa. Thomson's gazelles are pictured in the foreground, with scattered Acacia trees throughout. Photo: Brian Wilsey.

understory and some tree component. The proportion of tree cover in savannahs varies among continents but is usually considered to be less than 20 percent ground cover shaded by trees. A greater proportion of trees would lead to the site being classified as a woodland or forest. Scholes and Hall (1996) defined savannas as having 10–15 percent cover by woody plants, with a well-developed grassy layer; grasslands with <10 percent tree cover, and woodlands with 50–100 percent tree canopy cover. Forests have a complete canopy cover and usually have layers of woody vegetation (over, mid- and understories).

Much work has been conducted on how trees and grasses can coexist in savannah, and whether savannas are stable systems. Stability in this context is the ability to maintain the grass and scattered tree structure for multiple decades or more. Debates among scientists are ongoing about whether savannas are merely short-lived transition systems located between forest and grassland or are persistent systems that are worthy of study in their own right.

The way to answer the question of whether savannas are transitional or stable systems is to do studies that address two issues: 1) are there species that achieve their highest abundance in savannas, although they may by present in both grasslands and forests (Figure 8.9)? 2) Do grasses and trees utilize resources differently, which would enable the two growth forms to

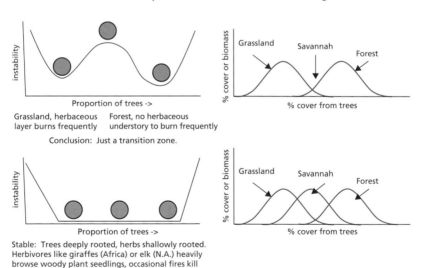

Figure 8.9   Contrasting views on savanna persistence and stability. The first (top panels) considers savanna as an unstable transition zone between forests and grasslands. The second (bottom panels) considers grasslands, savannas, and forests equally stable. Testing between these two views usually entails looking at the size of savannas, looking for species that have their peak abundance in savannas only (% cover or biomass), and looking for niche partitioning (usage of water at different soil depths or being consumed by different types of herbivores) between trees and herbaceous plants.

coexist indefinitely? Some people base their answer to this question on how extensive savannas are, with the assumption that a narrow band of savanna would be indicative a transitional (non-stable) system. At the wet end of the spectrum, savannahs grade into forests. However, using this logic, the huge grasslands in the interior of continents would be regarded as transitional systems between forests and deserts, which clearly is not the case.

Tropical savannahs are found in areas with intermediate rainfall and can become woodlands at higher rainfall amounts (Sankaran et al. 2005, Table 8.3). In Africa, browsing by giraffes and elephants keeps savannahs from converting to woodlands, especially in areas with >650 mm precipitation (Sankaran et al. 2005). Additional factors that permit stable coexistence is water partitioning

Table 8.3 Comparison of physical environments, species composition, and traits of dominant tree species in savannas versus forests (Ratnam et al. 2011).

Habitat type	Mesic savanna	Forest
Environmental descriptors	High-light understorey	Low-light understorey
	Frequently burnt	Fires rare, catastrophic
	Trees	Trees
Vegetation composition	Herbs	C_3 grasses
	C_4 grasses	Herbs
Adult trees	Relatively shorter	Relatively taller
Architecture	Narrower canopy diameter for a given basal area	Wider canopy diameter for a given basal area
Bark	Thick bark	Thin bark
	Lower specific leaf area	Higher specific leaf area
Canopy	Open crowns and higher light penetration through canopy	Dense crowns and lower light penetration through canopy
	Post-fire recovery of canopy either epicormic, or from protected apical buds	Limited post-fire recovery of canopy
Saplings	Many have vertical pole-like architecture	Varied, branched and unbranched architecture
	High root: shoot ratio	Low root: shoot ratio
	Large underground storage	Low underground storage
	Post-fire resprouting common under frequent, intense fires	Post-fire resprouting rare under frequent, intense fires
Seedlings	Rapid acquisition of resprouting ability through early allocation to root	No obvious acquisition of resprouting ability
	Persist through competition with C_4 grasses and repeated fire to sapling stage	Cannot persist through competition with grasses and repeated fires
Reproductive strategy of tree community	No or few species are obligate seeders, reproduction through root-suckering common	Reproduction through root-suckering uncommon

between shrubs or trees and grasses. Grasses tend to utilize water at surface soil layers, and trees or shrubs tend to utilize water at deeper soil layers (Holdo and Nippert 2015). Sankaran et al. (2005) found that savannahs are stable systems when precipitation is <650 mm. The cover of trees increases linearly with precipitation in this range. When precipitation is >650 mm, the savannahs are unstable and are maintained only by fire and herbivory. Ratnam et al. (2011) suggested that species traits can delineate a tropical savannah from a forest (Table 8.3). Tropical mesic savannahs are mixed tree-C_4 grass systems with fire-tolerant, and shade-intolerant species in their understories. Forests have little or no C_4 grass cover, and contain species that are fire intolerant and shade tolerant in their understories. Under this definition, many tropical dry forests would be considered to be savannahs.

There is much interest in North America in restoring savannas because of their rarity (Brudvig and Asbjornsen 2007). However, there are multiple human-caused ways that a grassy area with scattered trees can be created. A savanna-like area could have resulted from trees being planted into open grassland for shade for livestock in a former pasture, or trees could have been cleared and thinned to isolated individuals in a former forest. Relict savannahs in the Midwestern USA are identified by locating open grown oak trees and herbaceous species that may be present in forests or grasslands (Brudvig and Asbjornsen 2007). The abundance and persistence of some species are higher in savannas than in forests or grasslands, suggestive of savannas as a persistent system. Herbaceous species in North American Midwest that fit this description include Virginia wild rye (*Elymus virginicus*), horse gentian (*Triosteum perfoliatum*), and wild petunia (*Ruellia humilis*). Open grown oaks with thick bark that are fire resistant like bur oak (*Quercus macrocarpa*) in the north and black oak and post oak (*Quercus velutina* and *stellata*) in the south are also good indicators of former savannas.

9 Conclusions, Future Research Needs, and Issues

I expect that several issues will become increasingly important to study in grassland science. First, I expect continued interest in belowground ecology, especially as it pertains to carbon sequestration and release (Burke et al. 1989, Conant et al. 2001, Fierer and Jackson 2006, van der Heijden et al. 2008). Management for greater carbon sequestration in grasslands and savannas will continue to be emphasized by society. So far, empirical work is lagging behind modeling on this. Second, issues of sustainability will become more important as human populations near the 9 billion mark during mid-century (Jarchow and Liebman 2012). The amount of arable land is finite, and the land that we farm will have to become more productive to feed the growing human population. It will be a challenge to future grassland scientists to feed the world, while at the same time preventing further environmental degradation and biodiversity loss. Properly managing perennial grasslands to produce food, while at the same time preventing the soil erosion issues found in annual crop production (Schulte et al. 2017), provides an alternative form of production. Third, pollinating bees are in decline globally, and grasslands with abundant wildflower populations provide key pollen sources important to prevent their future decline (Baude et al. 2016). Finally, in the United States, the iconic butterfly species the Monarch butterfly is in decline. The decline is partially due to the use of glyphosate resistant crops, which has reduced the number of host plants in crop fields (Pleasants et al. 2017). Proper management of grassland sites to produce the monarch host plant milkweeds (*Asclepias* spp.) as well as companion plants that produce nectar along the migration route are needed. Current management plans recommend having 6 billion more milkweed stems planted in the central U.S. to save the milkweed, and many of these stems will have to be in grasslands (Pleasants et al. 2017). Properly managed grasslands have the potential to provide food, fuel, pollinator and butterfly habitat, and repositories for biodiversity.

The Biology of Grasslands. Brian J. Wilsey. Published 2018 by Oxford University Press. © Brian J. Wilsey 2018.
DOI 10.1093/oso/9780198744511.001.0001

A second major theme of future grassland science is likely to be in the fields of alternate states theory (Peters et al. 2004, Firn et al. 2010, Bestelmeyer et al. 2011), and restoration ecology. Some have suggested that the twenty-first century will be the century for restoration ecology. Many grasslands are degraded or have been replaced by exotic species, trees, or annual crop species. In many subhumid and humid grassland sites, few grasslands remain due to woody plant encroachment. It is poorly known if these forms of degradation are permanent (alternate states, Suding et al. 2004, Briske et al. 2005), or if they can be fully reversed with restoration that removes the exotic and woody plant species. For example, junipers (Chapter 8) have converted former diverse grasslands to low-diversity woodlands. A major assumption that many make about this conversion is that it can be easily be reversed. However, it is largely unknown how long the Junipers can be in place before it is not possible to restore the area back to diverse grassland. Many major grassland regions of the world may not have native extensive grasslands without active management in the future.

In many areas, the community that was found pre-settlement does not recover without major human inputs. These sites may form some kind of novel ecosystem that is different from what we have seen in the past. Novel ecosystems are likely to become more common as species move due to climate change (Wilsey et al. 2011), forming new species mixtures. Many interesting questions arise about novel grasslands, such as whether they differ from the native communities that they have replaced, how species interactions are altered, and if species are adapting to the new members of the community. Zuppinger et al. (2014) found rapid evolution in plant species from different mixture types, and it will be interesting to see how widespread this is as conditions change. These types of questions will likely be addressed by scientists in the future. Non-native species may be able to mimic the keystone species that are gone with proper management, for example cattle in place of bison, or extant giant tortoises in place of extinct giant tortoises. Grassland scientists in the future will have to determine if novel ecosystems represent a loss or an opportunity.

Bibliography

Abreu, R.C.R., Hoffmann, W.A., Vasconcelos, H.L., Pilon, N.A., Rossatto, D.R., and G. Durigan (2017). The biodiversity cost of carbon sequestration in tropical savanna. *Science Advances* 3: e1701284. doi:10.1126/sciadv.1701284.

Albertson, F.W. and Tomanek, G.W. (1965). Vegetation changes during a 30-year period in grassland communities near Hays, Kansas. *Ecology* 46: 714–20. doi:10.2307/1935011.

Alward, R.D., Detling, J.K., and D.G. Milchunas (1999). Grassland vegetation changes and nocturnal global warming. *Science* 283: 229–31.

Allen, C.R., Epperson, D.M., and A.S. Garmestani (2004). Red imported fire ant impacts on wildlife: A decade of research. *The American Midland Naturalist* 152: 88–103.

Allan, R.P. and B.J. Soden (2008). Atmospheric warming and the amplification of precipitation extremes. *Science* 321: 1481–4.

Allred B.W., S.D. Fuhlendorf, and R.G. Hamilton (2011). The role of herbivores in Great Plains conservation: comparative ecology of bison and cattle. *Ecosphere* 2(26). doi:10.1890/ES10-00152.1.

Alpert. 2006. The advantages and disadvantages of being introduced. Biological Invasions 8: 1523–34.

Ambrose, S.H. and DeNiro, M.J. (1986). The isotopic ecology of East African mammals *Oecologia* 69: 395–406.

Anderson, T.M., Ritchie, M.E., Mayemba, E., Eby, S., Grace, J.B., and S.J. McNaughton (2007). Forage nutritive quality in the Serengeti Ecosystem: The roles of fire and herbivory. *The American Naturalist* 170: 343–57.

Atjay, G.L., P. Ketner and P. Duvignead. (1979). Terrestrial primary production and phytomass. In B. Bolin, E.T. Degens, S. Kempe, and P. Ketner (eds.) *The Global Carbon Cycle*. Chichester: Wiley, pp. 129–81.

Atsatt, P.R. and D.J. O'Dowd (1976). Plant defense guilds. *Science* 193: 24–9.

Augustine, D.J. and B.W. Baker (2013). Associations of grassland bird communities with black-tailed prairie dogs in the North American Great Plains. *Conservation Biology* 27: 324–34.

Augustine, D.J., J.D. Derner, and D.G. Milchunas (2010). Prescribed fire, grazing, and herbaceous plant production in shortgrass steppe. *Rangeland Ecology and Management* 63: 317–23.

Augustine, D.J. and S.J. McNaughton (1998). Ungulate effects on the functional species composition of plant communities: Herbivore selectivity and plant tolerance. *Journal of Wildlife Management* 62: 1165–83.

Azpiroz, A.B., Isacch, J.P., Dias, R.A., Di Giacomo, A.S., Fontana, C.S., and Palarea, C.M. (2012). Ecology and conservation of grassland birds in southeastern South America: a review. *Journal of Field Ornithology* 83: 217–46. doi:10.1111/j.1557-9263.2012.00372.x.

Baer, S.G., Blair, J.M., Collins, S.L., and Knapp, A.K. (2004). Plant community responses to resource availability and heterogeneity during restoration. *Oecologia* 139: 617–29.

Baer, S.G., Kitchen, D.J., Blair, J.M., and Rice, C.W. (2002). Changes in ecosystem structure and function along a chronosequence of restored grasslands. *Ecological Applications* 12: 1688–701.

Bagchi, S. and Ritchie, M.E. (2010). Introduced grazers can restrict potential soil carbon sequestration through impacts on plant community composition. *Ecology Letters* 13(8): 959–68.

Bakker, J.P. and F. Berendse (2002). Constraints in the restoration of ecological diversity in grassland and heathland communities. *Trends in Ecology and Evolution* 14: 63–8.

Balchet, J.K., et al. (2013). Introduced annual grass increases regional fire activity across the arid western USA (1980–2009). *Global Change Biology* 19: 173–83.

Bardgett, R.D., Bowman, W.D., Kaufmann, R., and Schmidt, S.K. (2005). A temporal approach to linking aboveground and belowground ecology. *Trends in Ecology & Evolution* 20(11): 634–41.

Baude, M., Kunin, W.E., Boatman, N.D., Conyers, S., Davies, N., Gillespie, M.A.K., Morton, R.D., Smart, S.M., and J. Memmott (2016). Historical nectar assessment reveals the fall and rise of floral resources in Britain. *Nature* 530: 85–8.

Baxter, B.J.M, van Staden, J.E., Granger, N., and A.C. Brown (1994). Plant-derived smoke and smoke extracts stimulate seed germination of the fire-climax grass *Themeda triandra*. *Environmental and Experimental Botany* 34: 217–23.

Behmer, S.T. and A. Joern (2008). Coexisting generalist herbivores occupy unique nutritional feeding niches. *Proceedings of the National Academy of Sciences* 105: 1977–82.

Belsky, A.J. (1986). Does herbivory benefit plants? A review of the evidence. *American Naturalist* 127: 870–92.

Benson, E. and D. Hartnett (2006). The role of seed and vegetative reproduction in plant recruitment and demography in tallgrass prairie. *Plant Ecology* 187: 163–78.

Bertness. M.D. and R. Callaway (1994). Positive interactions in communities. *Trends in Ecology Evolution* 9: 191–3.

Beshta, R.L. and W.J. Ripple (2016). Riparian vegetation recovery in Yellowstone: The first two decades after wolf reintroduction. *Biological Conservation* 198: 93–103.

Bestelmeyer, B.T., D.P. Goolsby, and S.R. Archer (2011). Spatial perspectives in state-an-transition models: a missing link to land management? *Journal of Applied Ecology* 48: 746–57.

Biesmeijer, J.C., S.P.M. Roberts, M. Reemer, R. Ohlemuller, Edwards, M., et al. (2006). Parallel declines in pollinators and insect-pollinated plants in Britain and the Netherlands *Science* 313: 351–4.

Blair, J.M. (1997). Fire, N availability, and plant response in grasslands: a test of the transient maxima hypothesis. *Ecology* 78: 2359–68.

Blumenthal, D.M., Jordan, N.R., and Russelle, M.P. (2003). Soil carbon addition controls weeds and facilitates prairie restoration. *Ecological Applications* 13: 605–15.

Bond, W.J. (2008). What limits trees in C_4 grasslands and savannas? *Annual Review of Ecology, Evolution, and Systematics* 39: 641–59.

Bond, W.J. and J.E. Keeley (2005). Fire as a global 'herbivore': the ecology and evolution of flammable ecosystems. *Trends in Ecology and Evolution* 20: 387–94.

Bond, W.J. and Midgley, G.F. (2000). A proposed CO_2-controlled mechanism of woody plant invasion in grasslands and savannas. *Global Change Biology* 6: 865–9. doi:10.1046/j.1365-2486.2000.00365.x.

Borer, E.T., Seabloom, E.W., Gruner, D.S., and O'Halloran, L.R., et al. (2014). Herbivores and nutrients control grassland plant diversity via light limitation. *Nature* 508: 517–20.

Borer, E.T., Seabloom, E.W., Mitchell, C.E., and Cronin, J.P. (2014). Multiple nutrients and herbivores interact to govern diversity, productivity, composition, and infection in a successional grassland. *Oikos* 123(2): 214–24.

Bradshaw, A.D. (1987). Restoration: an acid test for ecology. In W.R. Jordan, M.E. Gilpin, and J.D. Aber (ed.) *Restoration Ecology: A Synthetic Approach*. Cambridge: Cambridge University Press, pp. 23–31.

Briggs, J.M, Knapp, A.K., Blair, J.M., Heisler, J.L., Hoch, G.A., Lett, M.S. and J.K. McCarron (2005). An ecosystem in transition: causes and consequences of the conversion of mesic grassland to shrubland. *Bioscience* 55: 243–54.

Briske, D.D., Fuhlendor, S.D., and F.E. Smeins (2005). State-and-transition models, thresholds, and rangeland health: A synthesis of ecological concepts and perspectives. *Rangeland Ecology & Management* 58: 1–10.

Brittingham, M.C. and S.A. Temple (1983). Have cowbirds caused forest songbirds to decline? *BioScience* 33: 31–5.

Brose, U., Martinez, N.D., and Williams, R.J. (2003). Estimating species richness: sensitivity to sample coverage and insensitivity to spatial patterns. *Ecology* 84(9). 2364–77.

Brown, J.S., Laundré, J.W., and M. Gurung. (1999). The ecology of fear: Optimal foraging, game theory, and trophic interactions. *Journal of Mammalogy* 80: 385–99.

Brudvig, L.A. and H. Asbjornsen. (2007). Stand structure, composition, and regeneration dynamics following removal of encroaching woody vegetation from Midwestern oak savannas. *Forest Ecology and Management* 244: 112–21.

Brye, K.R., Norman, J.M., and S.T. Gower (2002). Assessing the progress of a tallgrass prairie restoration in southern Wisconsin. *American Midland Naturalist* 148: 218–35.

Brye, K.R. and T.L. Riley (2009). Soil and plant property differences across a chronosequence of humid-temperate tallgrass prairie restorations. *Soil Science* 174: 346–57.

Burke, I.C., Lauenroth, W.K., and W.J. Parton. (1997). Regional and temporal variation in net primary production and nitrogen mineralization in grasslands. *Ecology* 78: 1330–40.

Burke, I.C., Yonker, C.M., Parton, W.J., Cole, C.V., Flach, K., and D.S. Schimel (1989). Texture, climate, and cultivation effects on soil organic matter content in U.S. grassland soils. *Soil Society America Journal* 53: 800–5.

Cadotte, M.W., Dinnage, R., and Tilman, D. (2012). Phylogenetic diversity promotes ecosystem stability. *Ecology* 93: S223–S233.

Callaway, R.M. and Ridenour, W.M. (2004). Novel weapons: invasive success and the evolution of increased competitive ability. *Frontiers in Ecology and the Environment* 2: 436–43.

Camill, P., McKone, M.J., Sturges, S.T., Severud, W.J., Ellis, E., Limmer, J., Martin, C.B., Navratil, R.T., Purdie, A.J., Sandel, B.S., Talukder, S., and Trout, A. (2004). Community- and ecosystem-level changes in species-rich tallgrass prairie restoration. *Ecological Applications* 14: 1680–94.

Cardinale, B.J., Gross, K., Fritschie, K., Flombaum, P., Fox, J., Rixen, C., van Ruijven, J., Reich, P., Scherer-Lorenzen, M., and B.J. Wilsey (2013). Can producer diversity simultaneously increase the productivity and stability of ecosystems? A meta-analysis of 34 experiments. *Ecology* 94: 1697–708.

Cardinale, B.J., Wright, J.P., Cadotte, M.W., Carroll, I.T., Hector, A., Srivastava, D.S., Loreau, M., and Weis, J.J. (2007). Impacts of plant diversity on biomass production increase through time because of species complementarity. *Proceedings of the National Academy of Sciences* 104(46): 18123–8.

Carter, D.L. and Blair, J.M. (2012). Recovery of native plant community characteristics on a chronosequence of restored prairies seeded into pastures in west-central Iowa. *Restoration Ecology* 20: 170–9. doi:10.1111/j.1526-100X.2010.00760.x.

Cerling, T.E., Harris, J.M., MacFadden, B.J., Leakey, M.G., Quade, J., Eisenmann, V., and J.R. Ehleringer (1997). Global vegetation change through the Miocene/Pliocene boundary. *Nature* 389: 153–8.

Chapin, F.S. III. (1980). The mineral nutrition of wild plants. *Annual Review Ecology Systematics* 11: 233–60.

Chapin, F.S. (1991). Integrated responses of plants to stress. *Bioscience* 41: 29–36.

Chapin, F.S. III, E.D. Schulze, and H.A. Mooney (1990). The ecology and economics of storage in plants. *Annual Review of Ecology and Systematics* 21: 423–47.

Chapin, F.S., E.S. Zavaleta, V.T. Eviner, R.L. Naylor, P.M. Vitousek, H L. Reynolds, D.U. Hooper, S. Lavorel, O.E. Sala, S.E. Hobbie, M.C. Mack, and S. Diaz (2000). Consequences of changing biodiversity. *Nature* 405: 234–42.

Checinska Sielaff, A., H.W. Polley, A. Fuentes-Ramirez, K. Hofmockel, and B.J. Wilsey (2017). Mycorrhizal colonization is higher, but impact on plants is lower in exotic than native plant species in Texas grasslands. Submitted to Biological Invasions.

Chesson, P. (2000). Mechanisms of maintenance of species diversity. *Annual Review of Ecology and Systematics* 31: 343–66.

Clauss, M., Hume, I.D., and J. Hummel (2010). Evolutionary adaptations of ruminants and their potential relevance for modern production systems. *Animal* 4: 979–92.

Cleland, E., Chiariello, N., Loarie, S., Mooney, H., and Field, C. (2006). Diverse responses of phenology to global changes in a grassland ecosystem. *Proceedings of the National Academy of Sciences of the United States of America* 103(37): 13740–4.

Clements, F.E. (1916). *Plant Succession: An Analysis of The Development of Vegetation.* Washington: Carnegie Institute.

Coley, P.D., Bryant, J.P., and Chapin, F.S. (1985). Resource availability and plant anti-herbivore defense. *Science* 230(4728): 895–9.

Collins, S.L and S.M. Glenn (1990). A hierarchical analysis of species' abundance patterns in grassland vegetation. *The American Naturalist* 135: 633–48.

Collins, S.L., Knapp, A.K., Briggs, J.M., Blair, J.M., and Steinauer, E.M. (1998). Modulation of diversity by grazing and mowing in native tallgrass prairie. *Science* 280: 745–7.

Conant, R.T., Paustian, K., and Elliott, E.T. (2001). Grassland management and conversion into grassland: effects on soil carbon. *Ecological Applications* 11: 343–55.

Connell, J.H. and R.O. Slatyer (1977). Mechanisms of succession in natural communities and their role in community stability and organization. *The American Naturalist* 111: 1119–44.

Cornwell, W.K., Cornelissen, J.H., Amatangelo, K., Dorrepaal, E., Eviner, V.T., Godoy, O., Hobbie, S.E., Hoorens, B., Kurokawa, H., Pérez-Harguindeguy, N., Quested, H.M., Santiago, L.S., Wardle, D.A., Wright, I.J., Aerts, R., Allison, S.D., Van Bodegom, P., Brovkin, V., Chatain, A., Callaghan, T.V., Díaz, S., Garnier, E., Gurvich, D.E., Kazakou, E., Klein, J.A., Read, J., Reich, P.B., Soudzilovskaia, N.A., Vaieretti, M.V., and Westoby, M. (2008). Plant species traits are the predominant control on litter decomposition rates within biomes worldwide. *Ecology Letters* 11(10): 1065–71.

Côté, S.D., Rooney, T.P., J.P. Trembley, C. Dussault, and D.M. Waller (2004). Ecological impacts of deer overabundance. *Annual Review of Ecology, Evolution and Systematics* 35: 113–47.

Coughenour, M.B. (1985). Graminoid responses to grazing by large herbivores: Adaptations, exaptations, and interacting processes. *Annals of Missouri Botanical Garden* 72: 852–63.

Craine, J.M., Ocheltree, T.W., Nippert, J.B., et al. (2013). Global diversity of drought tolerance and grassland climate-change resilience. *Nature Climate Change* 3: 63–7.

Crist, T.O. and Veech, J.A. (2006). Additive partitioning of rarefaction curves and species–area relationships: unifying α-, β-and γ-diversity with sample size and habitat area. *Ecology Letters* 9(8): 923–32.

Crist, T. O., Veech, J. A., Gering, J. C., and Summerville, K. S. (2003). Partitioning species diversity across landscapes and regions: a hierarchical analysis of α, β, and γ diversity. *The American Naturalist* 162: 734–43.

Daehler, C.C. (2003). Performance comparisons of co-occurring native and alien invasive plants: implications for conservation and restoration. *Annual Review of Ecology, Evolution and Systematics* 34: 183–211.

D'Antonio, C.M. and P.M. Vitousek (1992). Biological Invasions by Exotic Grasses, the Grass/Fire Cycle, and Global Change. *Annual Review of Ecology and Systematics* 23: 63–87.

de Dios Miranda, J., F.M. Padilla, and F.I. Pugnaire (2009). Response of a Mediterranean semiarid community to changing patterns of water supply. *Perspectives in Plant Ecology, Evolution and Systematics* 11: 255–66.

DeMalach, N., Zaady, E., and Kadmon, R. (2017). Contrasting effects of water and nutrient additions on grassland communities: A global meta-analysis. *Global Ecol Biogeogr* 26: 983–92.

de Wit, C.T., Tow, P.G., and Ennik, G.C. (1966). *Competition between legumes and grasses.* Wageningen: Pudoc. Centre for Agricultural Publications and Documentation Wageningen.

Derner, J.D., Lauenroth, W.K., Stapp, P., and D.J. Augustine (2009). Livestock as ecosystem engineers for grassland bird habitat in the Western Great Plains of North America. *Rangeland Ecology and Management* 62: 111–18.

Derner, J.D. and G.E. Schuman (2007). Carbon sequestration and rangelands: A synthesis of land management and precipitation effects. *Journal of Soil Water Conservation* 62: 77–85.

Detling, J.K. (1998). Mammalian herbivores: ecosystem-level effects in two grassland national parks. *Wildlife Society Bulletin* 26: 438–48.

Dickson T.L. and Gross K.L. (2015) Can the results of biodiversity-ecosystem productivity studies be translated to bioenergy production? *PLoS ONE* 10(9): e0135253. https://doi.org/10.1371/journal.pone.01352.

Dickson, T.L., Hopwood, J., and B.J. Wilsey (2012). Do priority effects benefit invasive plants more than native plants? An experiment with six grassland species. *Biological Invasions* 14: 2617–21.

Dickson, T.L., Mittelbach, G.G., Reynolds, H.L., and Gross, K.L. (2014). Height and clonality traits determine plant community responses to fertilization. *Ecology* 95: 2443–52. doi:10.1890/13-1875.1.

Dickson, T.L. and B.J. Wilsey (2009). Biodiversity and tallgrass prairie decomposition: the relative importance of species identity, evenness, richness and microtopography. *Plant Ecology* 201: 639–49.

Diggs, G.M., Lipscomb, B.L., and R.J. O'Kennon (1999). *Illustrated flora of North Central Texas.* Texas: Botanical Research Institute of Texas.

Dimitrakopoulos, P.G. and Schmid, B. (2004). Biodiversity effects increase linearly with biotope space. *Ecology Letters* 7(7): 574–83.

Dixon, K.W., Roche, S., and J.S. Pate (1995). The promotive effect of smoke derived from burnt native vegetation on seed germination of Western Australian plants. *Oecologia* 101: 185–92.

Donald, P.F., Sanderson, F.J., Burfield, I.J., and F.P.J. van Bommel (2006). Further evidence of continent-wide impacts of agricultural intensification on European farmland birds, 1990–2000. *Agriculture Ecosystems and Environment* 116: 189–96.

Dornbush, M.E. and Wilsey, B.J. (2010). Experimental manipulation of soil depth alters species richness and co-occurrence in restored tallgrass prairie. *Journal of Ecology* 98(1): 117–25.

Duffy, J.E. (2009). Why biodiversity is important to the functioning of real-world ecosystems. *Frontiers in Ecology and the Environment* 7(8): 437–44.

Dyer, M.I., Turner, C.L., and Seastedt, T.R. (1993). Herbivory and its consequences. *Ecological Applications* 3: 10–16.

Eby, S. and Ritchie, M.E. (2013). The impacts of burning on Thomson's gazelles', *Gazella thomsonii*, vigilance in Serengeti National Park, Tanzania. *African Journal of Ecology* 51: 337–42. doi:10.1111/aje.12044.

Edwards E.J., Osborne C.P., Strömberg C.A.E., and Smith S.A. (2010). The origins of C_4 grasslands: integrating evolutionary and ecosystem science. *Science* 328: 587–91.

Ehleringer, J.R., Cerling, T.E., and M.D. Dearing (2002). Atmospheric CO_2 as a global change driver influencing plant-animal interactions. *Integrative and Comparative Biology* 42: 424–30.

Ehleringer, J.R., T.E. Cerling, and B.R. Helliker (1997). C_4 photosynthesis, atmospheric CO_2, and climate. *Oecologia* 112: 285–99.

Ehrlich P. and Ehrlich A. (1981). *Extinction: The Causes and Consequences of the Disappearance of Species.* New York: Random House.

Epstein H.E., Lauenroth W.K., Burke I.C., Coffin D.P. (1997). Productivity patterns of C_3 and C_4 functional types in the U.S. Great Plains. *Ecology* 78: 722–31.

Eyheralde, P.G. (2015). *Bison-Mediated Seed Dispersal in a Tallgrass Prairie Reconstruction.* Dissertation, Iowa State University, Ames, IA, USA.

Falk, D.A., Palmer, M.A., and J.B. Zedler (2006). *Foundations of Restoration Ecology.* Washington: Island Press.

Fargione, J., C.S. Brown, and D. Tilman (2003). Community assembly and invasion: an experimental test of nuetral versus niche processes. *PNAS* 100: 8916–20.

Fay, P.A., Kaufman, D.M., Nippert, J.B., Carlisle, J.D., and C.W. Harper (2008). Changes in grassland ecosystem function due to extreme rainfall events: implications for responses to climate change. *Global Change Biology* 14: 1600–8.

Fay, P.A., et al. (2015). Grassland productivity limited by multiple nutrients. *Nature Plants* 1: 15080.

Fellbaum, C.R., Mensah, J.A., Cloos, A.J., Strahan, G.E., Pfeffer, P.E., Kiers, E.T., and Bücking, H. (2014). Fungal nutrient allocation in common mycorrhizal networks is regulated by the carbon source strength of individual host plants. *New Phytologist* 203(2): 646–56.

Field, C. and Mooney, H.A. (1986). The photosynthesis-nitrogen relationship in wild plants. In Givinsh T.J. (ed.) *On the Economy of Form and Function*. Cambridge: Cambridge University Press, pp. 25–55.

Fierer, N. and R.B. Jackson (2006). The diversity and biogeography of soil bacterial communities. *Proceedings of the National Academy of Sciences* 103: 626–31.

Fierer, N., J.W. Leff, B.J. Adams, U.N. Nielsen, S.T. Bates, C.L. Lauber, S. Owens, J.A. Gilbert, D.A. Wall, and J.G. Caporaso (2012). Cross-biome metagenomic analyses of soil microbial communities and their functional attributes. *Proceedings of the National Academy of Sciences* 109: 21390–5.

Firn, J., House, A.P.N., and Buckley, Y.M. (2010). Alternative states models provide an effective framework for invasive species control and restoration of native communities. *Journal of Applied Ecology* 47: 96–105.

Firn, J., Prober, S.M., and Y.M. Buckley (2012). Plastic traits of an exotic grass contribute to its abundance but are not always favourable. *PlosOne* doi.org/10.1371/journal.pone.0035870

Fitter, A.H. and R.S.R. Fitter (2002). Rapid changes in flowering time in British plants. *Science* 296: 1689–91.

Ford, A.T., Goheen, J.R., Otieno, T.O., Bidner, L., Isbell, L.A., Palmer, T.M., Ward, D., Woodroffe, R., and R.M. Pringle (2014). Large carnivores make savanna tree communities less thorny. *Science* 346: 346–9.

Ford, P.L. and G.V. Johnson (2006). Effects of dormant—vs. growing-season fire in shortgrass steppe: biological soil crust and perennial grass responses. *Journal of Arid Environments* 67: 1–14.

Foster, B.L., Dickson, T.L., Murphy, C.A., Karel, I.S., and V.H. Smith (2004). Propagule pools mediate community assembly and diversity-ecosystem regulation along a grassland productivity gradient. *Journal of Ecology* 92: 435–49. doi:10.1111/j.0022-0477.2004.00882.x.

Foster, B.L., Kindscher, K., G.R. Houseman, and C.A. Murphy (2009). Effects of hay management and native species sowing on grassland community structure, biomass, and restoration. *Ecological Applications* 19: 1884–96.

Fox, J.W. (2005). Interpreting the 'selection effect' of biodiversity on ecosystem function. *Ecology Letters* 8(8): 846–56.

Frank, D. and McNaughton, S. (1992). The Ecology of Plants, Large Mammalian Herbivores, and Drought in Yellowstone National Park. *Ecology* 73(6): 2043–58.

Frank, D.A., R.L. Wallen, and P.J. White (2016). Ungulate control of grassland production: grazing intensity and ungulate species composition in Yellowstone Park. *Ecosphere* 7(11): e01603. 10.1002/ecs2.1603.

Fretwell, S.D. (1977). Regulation of plant communities by the food chains exploiting them. *Perspectives Biology and Medicine* 20: 169–85.

Fridley, J.D. (2012). Extended leaf phenology and the autumn niche in deciduous forest invasions. *Nature* 485: 359–62.

Froyd, C.A., et al. (2013). The ecological consequences of megafaunal loss: giant tortoises and wetland biodiversity. *Ecology Letters* 17: 144–54.

Fuhlendorf, S.D. and D.M. Engle. (2001). Restoring heterogeneity on rangelands: ecosystem management based on evolutionary grazing patterns. *Bioscience* 51: 625–32.

Fuhlendorf, S.D., Harrell, W.C., Engle, D.M., Hamilton, R.G., Davis, C.A., and Leslie, D.M. (2006). Should heterogeneity be the basis for conservation? Grassland bird response to fire and grazing. *Ecological Applications* 16: 1706–16.

Fukami, T. (2015). Historical contingency in community Assembly: Integrating niches, species pools, and priority effects. *Annual Review of Ecology, Evolution, and Systematics* 46: 1–23.

Fukami, T. and M. Nakajima (2011). Community assembly: alternative stable states or alternative transient states? *Ecology Letters* 14: 973–84.

Funk, J.L., Cleland, E.E., Suding, K.N., and Zavaleta, E.S. (2008). Restoration through reassembly: plant traits and invasion resistance. *Trends in Ecology & Evolution* 23: 695–703.

Gallagher, R.V. (2016). Correlates of range size variation in the Australian seed-plant flora. *Journal of Biogeography* 43: 1287–98. doi:10.1111/jbi.12711.

Gallinat, A.S., Primack, R.B., and D.L. Wagner (2015). Autumn, the neglected season in climate change research. *Trends in Ecology and Evolution* 30: 169–76.

Gaston, K.J. (2010). Valuing common species. *Science*, 327(5962): 154–5.

Gaudet, C.L. and P.A. Keddy (1988). A comparative approach to predicting competitive ability from plant traits. *Nature* 334: 242–3.

Gessner, M.O., Swan, C.M., Dang, C.K., McKie, B.G., Bardgett, R.D., Wall, D.H., and Hättenschwiler, S. (2010). Diversity meets decomposition. *Trends in Ecology & Evolution* 25(6): 372–80.

Gibson, D.J. (2009). *Grasses and Grassland Ecology*. Oxford: Oxford University Press.

Gill, R.A., H.W. Polley, H.B. Johnson, L.J. Anderson, H. Mahaereli, and R.B. Jackson (2002). Nonlinear grassland responses to past and future atmospheric CO_2. *Nature* 417: 274–82.

Givnish, T.J. (1994). Does diversity beget stability? *Nature* 371: 113–14.

Gleason, H.A. (1922). The vegetational history of the middle west. *Annals of the Association of American Geographers* 12: 39–85.

Gordon, I.J. and A.W. Illius. (1994). The functional significance of the browser-grazer dichotomy in African ruminants. *Oecologia* 98: 167–75.

Gotelli, N.J. and Arnett, A.E. (2000). Biogeographic effects of red fire ant invasion. *Ecology Letters* 3: 257–61. doi:10.1046/j.1461-0248.2000.00138.x.

Grace, J.B., Michael Anderson, T., Smith, M.D., Seabloom, E., Andelman, S.J., Meche, G., Weiher, E., Allain, L.K., Jutila, H., Sankaran, M., Knops, J., Ritchie, M., and Willig, M.R. (2007). Does species diversity limit productivity in natural grassland communities? *Ecology Letters* 10(8): 680–9.

Gross, K. and B. Cardinale (2007). Does species richness drive community production or vice versa? Reconciling historical and contemporary paradigms in competitive communities. *The American Naturalist* 170: 207–20.

Gravuer, K., Sullivan, J.J., Williams, P.A. and R.P. Duncan (2008). Strong human association with plant invasion success for *Trifolium* introductions to New Zealand. *PNAS* 105:6344–9.

Grime, J. (1977). Evidence for the existence of three primary strategies in plants and its relevance to ecological and evolutionary theory. *The American Naturalist* 111(982): 1169–94.

Grime, J.P. (1998). Benefits of plant diversity to ecosystems: immediate, filter and founder effects. *Journal of Ecology* 86: 902–10. doi:10.1046/j.1365-2745.1998.00306.x.

Grime, J.P., Thompson, K, Hunt, R., Hodgson, J.G., and J.H.C. Cornelissen (1997). Integrated screening validates primary axes of specialisation in plants. *Oikos* 79: 259–81.

Grime, J.P., Mackey, J.M.L., Hillier, S.H., and D.J. Read (1987). Floristic diversity in a model system using experimental microcosms. *Nature* 328: 420–2.

Grman E., Bassett, T., and Brudvig, L.A. (2013). Confronting contingency in restoration: management and site history determine outcomes of assembling prairies, but site characteristics and landscape context have little effect. *Journal of Applied Ecology*. doi: 10.1111/1365-2664.12135.

Grman, E. and Suding, K.N. (2010). Within-year soil legacies contribute to strong priority effects of exotics on native California grassland communities. *Restoration Ecology* 18(5): 664–70.

Guerrero-Ramirez, N.R., Craven, Reich, P.B., Ewel, J.J., Isbell, F., Koricheva, J., Parrotta, J.A., Auge, H. Erickson, H.E., Forrester, D.I. Hector, A. Joshi, J. Montagnini, F. Palmborg, C., Piotto, D. Potvin, C., Roscher, C., van Ruijven, J. Tilman, D., Wilsey, B., and N. Eisenhauer (2017). Temporal divergence of ecosystem functioning among plant diversity levels in grassland and forest experimental ecosystems. *Nature Ecology and Evolution* in press.

Hagerman, A.E. and C.T. Robbins (1993). Specificity of tannin-binding salivary proteins relative to diet selection by mammals. *Canadian Journal of Zoology* 71: 628–63.

Handa, I. T., Aerts, R., Berendse, F., Berg, M. P., Bruder, A., Butenschoen, O., Chauvet, E., Gessner, M.O., Jabiol, J., Makkonen, M., McKie, B. G., Malmqvist, B., Peeters, E.T.H.M., Scheu, S., Schmid, B., van Ruijven, J., Vos, V.C.A., and Hättenschwiler, S. (2014). Consequences of biodiversity loss for litter decomposition across biomes. *Nature* 509(7499): 218–21.

Hansen, M.J. and Gibson, D.J. (2014). Use of multiple criteria in an ecological assessment of a prairie restoration chronosequence. *Applied Vegetation Science* 17: 63–73. doi:10.1111/avsc.12051.

Harper, J.L. (1961). Approaches to the study of plant competition. *Symposia of the Society for Experimental Biology* 15: 1–39.

Harpole, W.S., Sullivan, L.L., et al. (2016). Addition of multiple limiting resources reduces grassland diversity. *Nature* 537: 93–6.

Harpole, W.S. and D. Tilman (2006). Non-neutral patterns of species abundance in grassland communities. *Ecology Letters* 9:15–23. doi:10.1111/j.1461-0248.2005.00836.x.

Harrison, S., Ross, S.J., and J.H. Lawton (1992). Beta diversity on geographic gradients in Britain. *Journal of Animal Ecology* 61: 151–8.

Harshman, J., Braun, E.L., Braun, M.J., Huddleston, C.J., Bowie, R.C., Chojnowski, J.L., Hackett, S.J., Han, K., Kimball, R.T., Marks B.D., Miglia, K. J., Moore, W.S., Reddy, S., Sheldon, F.H., Steadman, D.W., Steppan, S.J., Witt, C.C., and Yuri, T. (2008). Phylogenomic evidence for multiple losses of flight in ratite birds. *Proceedings of the National Academy of Sciences* 105(36): 13462–7.

Haukioja, E. (1991). Induction of defenses in trees. *Annual Review Entomology* 36: 25–42.

Hautier, Y., Seabloom, E.W., Borer, E.T., Adler, P.B., Harpole, W.S., Hillebrand, H., Lind, E.M., MacDougall, A.S., Stevens, C.J., Bakker, J.D., Buckley, Y. M., Chu, C., Collins, S.L., Daleo, P., Damschen, E.I., Davies, K.F., Fay, P.A., Firn, J., Gruner, D.S., Jin, V.L., Klein, J.A., Knops, J.M.H., La Pierre, K.J., Li, W., McCulley, R.L., Melbourne, B.A., Moore, J.L., O'Halloran, L.R., Prober, S.M., Risch, A.C., Sankaran, M., Schuetz, M., and Hector, A. (2014). Eutrophication weakens stabilizing effects of diversity in natural grasslands. *Nature* 508(7497): 521–5.

Hawkes, C.V. and J.J. Sullivan (2001). The impact of herbivory on plants in different resource conditions: a meta-analysis. *Ecology* 82: 2045–58.

Hector, A. and Bagchi, R. (2007). Biodiversity and ecosystem multifunctionality. *Nature* 448(7150): 188–90.

Hector, A., Hautier, Y., Saner, P., Wacker, L., Bagchi, R., Joshi, J., Scherer-Lorenzen, M., Spehn, E.M., Bazeley-White, E., Weilenmann, M., Caldeira, M.C., Dimitrakopoulos, P.G., Finn, J.A., Huss-Danell, K., Jumpponen, A., Mulder, C.P.H., Palmborg, C., Pereira, J.S., Siamantziouras, A.S.D., Terry, A.C., Troumbis, A.Y., Schmid, B., and Loreau, M. (2010). General stabilizing effects of plant diversity on grassland productivity through population asynchrony and overyielding. *Ecology* 91: 2213–20.

Hector, A., B. Schmid, C. Beierkuhnlein, M.C. Caldeira, M. Diemer, P.G. Dimitrakopoulos, J. A. Finn, H. Freitas, P.S. Giller, J. Good, R. Harris, P. Högberg, K. Huss-Danell, J. Joshi, A. Jumpponen, C. Körner, P. W. Leadley, M. Loreau, A. Minns, C.P.H. Mulder, G. O'Donovan, S.J. Otway, J.S. Pereira, A. Prinz, D.J. Read, M. Scherer-Lorenzen, E.D. Schulze, A.S.D. Siamantziouras, E.M. Spehn, A.C. Terry, A.Y. Troumbis, F.I. Woodward, S. Yachi, J.H. Lawton. (1999). Plant Diversity and Productivity Experiments in European Grasslands. Science 286: 1123–27

Heisler-White, J.L., Blair, J.M., Kelly, E.F., Harmony, K., and A.K. Knapp (2009). Contingent productivity responses to more extreme rainfall regimes across a grassland biome. *Global Change Biology* 15: 2844–904.

Hejcman, M., Hejcmanová, P., Pavlů, V., and Beneš, J. (2013). Origin and history of grasslands in Central Europe—a review. *Grass Forage Science* 68: 345–63. doi:10.1111/gfs.12066.

Hillebrand, H., Bennett, D.M., and Cadotte, M.W. (2008). Consequences of dominance: a review of evenness effects on local and regional ecosystem processes. *Ecology* 89(6): 1510–20.

Hobbs, N.T. (1996). Modification of ecosystems by ungulates. *Journal of Wildlife Management* 60: 695–713.

Hobbs, R.J., Arico, S., Aronson, J., Baron, J.S., Bridgewater, P., Cramer V.A., Epstein, P.R. Ewel, J.J., Klink, C.A., Lugo, A.E., Norton, D., Ojima, D., Richardson, D.M., Sanderson, E.W., Valladeres, F., Vilá, M, Zomoro, R., and M. Zobel (2006). Novel ecosystems: theoretical and management aspects of the new ecological world order. *Global Ecology and Biogeography* 15: 1–7.

Hofmann, R.R. (1973). *The Ruminant Stomach: Stomach Structure and Feeding Habits of East African Game Ruminants*. Nairobi, Kenya: East African Literature Bureau.

Holdo, R.M. and Nippert, J.B. (2015). Transpiration dynamics support resource partitioning in African savanna trees and grasses. *Ecology* 96: 1466–72. doi:10.1890/14-1986.1

Hooper, D.U., Chapin, F.S., Ewel, J.J., Hector, A., Inchausti, P., Lavorel, S., Lawton, J.H., Lodge, D.M., Loreau, M., Naeem, S., Schmid, B., Setälä, H., Symstad, A.J., Vandermeer, J., and Wardle, D.A. (2005). Effects of biodiversity on ecosystem functioning: a consensus of current knowledge. *Ecological Monographs* 75(1): 3–35.

Horncastle, V.J., Hellgren, E.C., Mayer, P.M., Ganguli, A.C., Engle, D.M., and D.M. Leslie (2005). Implications of invasion by *Juniperus virginiana* on small mammals in the southern Great Plains. *Journal of Mammalogy* 86: 1144–55.

Howe, H. (1994). Response of early- and late-flowering plants to fire season in experimental prairies. *Ecological Application* 4: 121–33.

Howe, H. (2011). Fire season and prairie forb richness in a 21-y experiment. *Ecoscience* 18: 317–28.

Howe, H.F., Brown, J.S., and Zorn-Arnold, B. (2002). A rodent plague on prairie diversity. *Ecology Letters* 5: 30–36. doi:10.1046/j.1461-0248.2002.00276.x.

Hubbell, S.P. (2005). Neutral theory in community ecology and the hypothesis of functional equivalence. *Ecology* 19: 166–72.

Huston, M.A. (1997). Hidden treatments in ecological experiments: re-evaluating the ecosystem function of biodiversity. *Oecologia* 110(4): 449–60.

Huxman, T.E., Smith, M.D., Fay, P.A., Knapp, A.K., Shaw, A.R., et al. (2004). Convergence across biomes to a common rain-use efficiency. *Nature* 429: 651–4.

Illius, A.W. and I.J. Gordon (1992). Modelling the nutritional ecology of ungulate herbivores: evolution of body size and competitive interactions. *Oecologia* 89: 428–34.

Inouye, D.W. (2008). Effects of climate change on phenology, frost damage, and floral abundance of montane wildflowers. *Ecology* 89: 353–62.

Isbell, F., Calcagno, V., Hector, A., Connolly, J., Harpole, W.S., Reich, P.B., Scherer-Lorenzen, M., Schmid, B., Tilman, D., van Ruijven, J., Weigelt, A., Wilsey, B.J., Zavaleta, E.S., and M. Loreau (2011). High plant diversity is needed to maintain ecosystem services. *Nature* 477: 199–202.

Isbell, F., Craven, D., Connolly, J., Loreau, M., Schmid, B., Beierkuhnlein, C., Bezemer, T.M., Bonin, C., Bruelheide, H., Luca, E., Ebeling, A., Griffin, J.N., Guo, Q., Hautier, Y., Hector, A., Jentsch, A., Kreyling, J., Lanta, V., Manning, P., Meyer, S.T., Mori, A.S., Naeem, S., Niklaus, P.A., Polley, H.W., Reich, P.B., Roscher, C., Seabloom, E.W., Smith, M.D., Thakur, M.P., Tilman, D., Tracy, B.F., van der Putten, W.H., van Ruijven, J., Weigelt, A., Weisser, W.W., Wilscy, B., and Eisenhauer, N. (2015). Biodiversity increases the resistance of ecosystem productivity to climate extremes. *Nature* 526: 574–7.

Isbell, F.I., Losure, D.A., Yurkonis, K.A., and Wilsey, B.J. (2008). Diversity–productivity relationships in two ecologically realistic rarity–extinction scenarios. *Oikos* 117(7): 996–1005.

Isbell, F.I., Polley, H.W., and Wilsey, B.J. (2009). Biodiversity, productivity and the temporal stability of productivity: patterns and processes. *Ecology Letters* 12(5): 443–51.

Isbell, F., Tilman, D., Polasky, S., Binder, S., and Hawthorne, P. (2013). Low biodiversity state persists two decades after cessation of nutrient enrichment. *Ecology Letters* 16(4): 454–60.

Isbell, F.I. and B.J. Wilsey (2011). Increasing native, but not exotic, biodiversity enhances ecosystem functioning in ungrazed and intensely grazed grasslands. *Oecologia* 165: 771–81.

Jackson, R.D., Paine, L.K., and J.E. Woodis (2010). Persistence of native C_4 grasses under high-intensity, short-duration summer bison grazing in the eastern tallgrass prairie. *Restoration Ecology* 18: 65–73.

Jaramillo, V.J. and J.K. Detling (1992). Small-scale heterogeneity in a semi-arid North American grassland. II. Cattle grazing of simulated urine patches. *Journal of Applied Ecology* 29: 9–13.

Jarchow, M.E. and M. Liebman (2012). Nutrient enrichment reduces complementarity and increases priority effects in prairies managed for bioenergy. *Biomass and Bioenergy* 36: 381–9.

Jefferson, L.V., Pennacchio, M., Havens, K., Forsberg, B., and D. Sollenberger (2008). *Ex situ* germination responses of midwestern USA prairie species to plant-derived smoke. *American Midland Naturalist* 159: 251–6.

Jenny, H. (1980). The Soil Resource. *Ecological Studies 37*. New York, Heidelberg, Berlin: Springer-Verlag.

Joern, A. (2005). Disturbance by fire frequency and bison grazing modulate grasshopper assemblages in tallgrass prairie. *Ecology* 86: 861–73.

Johansen, J.R. (1993). Cryptogamic crusts of semiarid and arid lands of North America. *Journal of Phycology* 29: 140–7.

Jost, L. (2007). Partitioning diversity into independent alpha and beta components. *Ecology* 88: 2427–39.

Kabigumila, J. (2001). Sighting frequency and food habits of the leopard tortoise, *Geochelone pardalis*, in northern Tanzania. *African Journal of Ecology* 39: 276–85.

Kang, L., X. Han, Z. Zhang, and O.J. Sun (2007). Grassland ecosystems in China: review of current knowledge and research advancement. Proceedings of the Royal Society B. doi:10.1098/rstb.2007.2029.

Karl, T.R. and R.W. Knight (1998). Secular trends of precipitation amount, frequency, and intensity in the United States. *Bulletin of the American Meteorological Society* 362(1482). doi.org/10.1175/1520-0477.

Keever, C. (1950). Causes of succession on old fields of the piedmont, North Carolina. *Ecological Monographs* 20: 229–50. doi:10.2307/1948582.

Kiers, E.T., Duhamel, M., Beesetty, Y., Mensah, J.A., Franken, O., Verbruggen, E., Fellbaum, C.R., Kowalchuk, G.A., Hart, M.M., Bago, A., Palmer, T.M., West, S.A., Vandenkoornhuyse, P., Jansa J., and Bucking, H. (2011). Reciprocal rewards stabilize cooperation in the mycorrhizal symbiosis. *Science* 333: 880–2.

Kindscher, K. and P.V. Wells (1995). Prairie plant guilds: A multivariate analysis of prairie species based on ecological and morphological traits. *Vegetatio* 117: 29–50.

Kirwan, L., Lüscher, A., Sebastia, M. T., Finn, J. A., Collins, R. P., Porqueddu, C., Heldgadottir, A., Baadshaug, O.H., Brophy, C., Coran, C., Dalmannsdottir, S., Delgado, I., Elgersma, A., Fothergill, M., Frankow-Lindberg, B.E., Golinski, P., Grieu, P., Gustavsson, A.M., Höglind, M., Huguenin-Elie, O., Iliadis, C., Jørgensen, M., Kadziuliene, Z., Karyotis, T., Lunnan, T., Malengier, M., Maltoni, S., Meyer, V., Nyfeler, D.,Nykanen-Kurki, P., Parente, J., Smit, H.J., Thumm, U., and Connolly, J. (2007). Evenness drives consistent diversity effects in intensive grassland systems across 28 European sites. *Journal of Ecology* 95(3): 530–9.

Kittelson, P.M., Wagenius, S., Nielsen, R., Qazi, S., Howe, M., Kiefer, G. and Shaw, R.G. (2015). How functional traits, herbivory, and genetic diversity interact in *Echinacea*: implications for fragmented populations. *Ecology* 96: 1877–86. doi:10.1890/14-1687.1.

Klaas, B.A., Danielson, B.J., and K.A. Moloney (1998). Influence of pocket gophers on meadow voles in a tallgrass prairie. *Journal of Mammalogy* 79: 942–52.

Klironomos, J.N. (2003). Variation in plant response to native and exotic arbuscular mycorrhizal fungi. *Ecology* 84: 2292–301.

Knapp, A., Blair, J., Briggs, J., Collins, S., Hartnett, D., Johnson, L., and Towne, E. (1999). The keystone role of bison in North American tallgrass prairie: bison increase habitat heterogeneity and alter a broad array of plant, community, and ecosystem processes. *BioScience* 49(1): 39–50.

Knapp, A.K., Fay, P.A., Blair, J.M., Collins, S.L., Smith, M.D., Carlisle, J.D., Harper, C.W., Danner, B.T., Lett, M.S., and McCarron, J.K. (2002). Rainfall variability, carbon cycling, and plant species diversity in a mesic grassland. *Science* 298: 2202–5.

Knapp, A.K., McCarron, J.K., Silleti, G.A., Hoch, G.I., Heisler, M.S., Lett, J.M., Blair, J.M, Briggs, J.M., and M.D. Smith (2008). Ecological consequences of the replacement of native grassland by *Juniperus virginiana* and other woody plants. In O.W. Van Auken (ed.) *Western North American Juniperus Communities: A Dynamic Vegetation Type*. New York: Springer, pp. 156–69.

Knapp, A.K. and T.R. Seastedt (1986). Detritus accumulation limits productivity of tallgrass prairie. *Bioscience* 36: 662–8.

Knapp, A.K. and M.D. Smith (2001). Variation among biomes in temporal dynamics of aboveground primary production. *Science* 291: 481–4.

Knops, J.M.H., Wedin, D., and D. Tilman (2001). Biodiversity and Decomposition in Experimental Grassland Ecosystems. *Oecologia* 126: 429–33.

Kucharik C.J., Fayram, N.J., and K.N. Cahill (2006). A paired study of prairie carbon stocks, fluxes, and phenology: comparing the world's oldest prairie restoration with an adjacent remnant. *Global Change Biology* 12: 122–39.

Kuchler, A.W. (1964). *Potential Natural Vegetation of the Conterminous United States*. American Geographical Society, Special Publication No. 36.

Kulmatiski, A. (2006). Exotic plants establish persistent communities. *Plant Ecology* 187: 261–75.

Kulmatiski, A. and K.H. Beard (2013). Woody plant encroachment facilitated by increased precipitation intensity. *Nature Climate Change* 3: 833–7.

Lande, R. (1996). Statistics and partitioning of species diversity, and similarity among multiple communities. *Oikos* 71: 5–13.

Larsen, T.H., Williams, N.M., and Kremen, C. (2005). Extinction order and altered community structure rapidly disrupt ecosystem functioning. *Ecology Letters* 8(5): 538–47.

Lauenroth, W.K., I.C. Burke, and M.P. Gutmann (1999). The structure and function of ecosystems in the central North American grassland region. *Great Plains Research* 9: 223–59.

Lawton, J.H. (1994). What do species do in ecosystems? *Oikos* 71: 367–74.

Leach, M.K. and T.J. Givnish (1996). Ecological determinants of species loss in remnant prairies. *Science* 273: 1555–8.

Lepš, J. (2004). Variability in population and community biomass in a grassland community affected by environmental productivity and diversity. *Oikos* 107(1): 64–71.

Levi, T. and Wilmers, C.C. (2012). Wolves–coyotes–foxes: a cascade among carnivores. *Ecology* 93: 921–9.

Liao, C. Peng, R, Luo, Y., Zhou, X., Wu, X., Fang, C., Chen, J., and B. Li (2008). Altered ecosystem carbon and nitrogen cycles by plant invasion: a meta-analysis. *New Phytologist* 177: 706–14.

Lieth, H. (1975). Primary production of the major vegetation units of the world. Primary Productivity of the Biosphere. In *Ecol. Studies, 14*. H. Lieth, H. and R.H. Whittaker (eds.) Berlin, Heidleberg: Springer-Verlag, pp. 203–5.

Lincoln, D.E., Fajer, E.D. and R.H. Johnson (1993). Plant-insect herbivore interactions in elevated CO_2 environments. *Trends in Ecology and Evolution* 8: 64–8.

Liu, J., Feng, C., Wang, D., Wang, L., Wilsey, B.J., and Z. Zhong (2015). Impacts of grazing by different large herbivores in grassland depend on plant species diversity. *J. Applied Ecology* 52: 1053–62.

Liu, H., Han, X., Li, L., Huang, J., Liu, H., and X. Li (2009). Grazing density effects on cover, species composition, and nitrogen fixation of biological soil crust in an Inner Mongolia steppe. *Rangeland Ecology and Management* 62: 321–7.

Loreau, M. (2000). Are communities saturated? On the relationship between α, β and γ diversity. *Ecology Letters* 3(2): 73–6.

Loreau, M., and Hector, A. (2001). Partitioning selection and complementarity in biodiversity experiments. *Nature*, 412(6842): 72–6.

Loreau, M. and de Mazancourt, C. (2008). Species synchrony and its drivers: neutral and nonneutral community dynamics in fluctuating environments. *The American Naturalist* 172: E48–E66.

Ma, M. (2005). Species richness vs evenness: independent relationship and different responses to edaphic factors. *Oikos* 111: 192–8.

MacDougal, A.S. and R. Turkington (2005). Are invasive species the drivers or the passengers of change in degraded ecosystems? *Ecology* 86: 42–56.

Mack, R.N. and Lonsdale, W.M. (2001). Humans as global plant dispersers: getting more than we bargained for. *BioScience* 51: 95–102.

Marshall, K.N., Hobbs, N.T., and Cooper, D.J. (2013). Stream hydrology limits recovery of riparian ecosystems after wolf reintroduction. *Proceedings of the Royal Society B* 280: 20122977. http://dx.doi.org/10.1098/rspb.2012.2977.

Martin, L.M., Harris, M., and B.J. Wilsey (2015). Phenology and temporal niche overlap differ between novel, exotic- and native-dominated grasslands for plants, but not for pollinators. *Biological Invasions* 17: 2633–44.

Martin, L.M., Moloney, K.A., and B.J. Wilsey (2005). An assessment of grassland restoration success using species diversity components. *Journal of Applied Ecology* 42: 327–36.

Martin, L.M., Polley, H.W., Daneshgar, P.P., Harris, M.A., and B.J. Wilsey (2014). Biodiversity, photosynthetic mode, and ecosystem services differ between native and novel ecosystems. *Oecologia* 175: 687–97.

Martin, L.M. and B.J. Wilsey (2006). Assessing grassland restoration success: relative roles of seed additions and native ungulate activities. *Journal of Applied Ecology* 43: 1098–110.

Martin, L.M. and B.J. Wilsey (2012). Assembly history alters alpha and beta diversity, exotic–native proportions and functioning of restored prairie plant communities. *Journal of Applied Ecology* 49: 1436–45.

Martin, L.M. and B.J. Wilsey (2014). Native-species seed additions do not shift restored prairie plant communities from exotic to native states. *Basic and Applied Ecology* 15: 297–304.

Martin, L.M. and B.J. Wilsey (2015). Novel, exotic-dominated grasslands exhibit altered patterns of beta diversity relative to native grasslands. *Ecology* 96: 1042–51.

Massey, F.P. and Hartley, S.E. (2009). Physical defences wear you down: progressive and irreversible impacts of silica on insect herbivores. *Journal of Animal Ecology* 78: 281–91.

Matamala, R., Jastrow, J.D., Miller, R.M., and Garten, C.T. (2008). Temporal changes in C and N stocks of restored prairie: implications for C sequestration strategies. *Ecological Applications* 18: 1470–1488. doi:10.1890/07-1609.1.

Matthews, E. (1983). Global vegetation and land use: New high-resolution data bases for climate studies. *Journal of Climate and Applied Meteorology* 22: 474–87.

McArthur, R.H. and E.O. Wilson (1967). *The Theory of Island Biogeography*. Princeton: Princeton University Press.

McGranahan, D.A., Engle, D.M., Fuhlendorf, S.D., Miller, J.R., and Debinski, D.M. (2012). An invasive cool-season grass complicates prescribed fire management in a native warm-season grassland. *Natural Areas Journal* 32: 208–14.

McIvor, J.G. (2005). Australian grasslands. In *Grasslands of the World*. Chapter 9 in Plant Production and Protection Series No. 34.

McLachlan, S.M. and A L. Knispel (2008). Assessment of long-term tallgrass prairie restoration in Manitoba, Canada. *Biological Conservation* 124: 75–88.

McNaughton, S.J. (1977). Diversity and stability of ecological communities: a comment on the role of empiricism in ecology. *American Naturalist* 111(979): 515–25.

McNaughton, S. J. (1979). Grazing as an optimization process: Grass-ungulate relationships in the Serengeti. *The American Naturalist* 113(5): 691–703.

McNaughton, S.J. (1983). Serengeti grassland ecology: the role of composite environmental factors and contingency in community organization. *Ecological monographs* 53(3): 291–320.

McNaughton, S.J. (1984). Grazing lawns: Animals in herds, plant form, and coevolution. *The American Naturalist* 124(6): 863–86.

McNaughton, S.J. (1985). Ecology of a grazing ecosystem: the Serengeti. *Ecological Monographs* 55: 259–94.

McNaughton, S.J. (1986). On plants and herbivores. *The American Naturalist* 128: 765–70.

McNaughton, S.J. (1988). Mineral nutrition and spatial concentrations of African ungulates. *Nature* 334: 343–5.

McNaughton, S.J., F.F. Banyikwa, and M.M. McNaughton (1997). Promotion of the Cycling of Diet-Enhancing Nutrients by African Grazers. *Science* 278: 1798–800.

McNaughton, S.J., Banyikwa, F.F., and McNaughton, M.M. (1998). Root biomass and productivity in a grazing ecosystem: the Serengeti. *Ecology* 79: 587–92.

McNaughton, S.J., Milchunas, D.G., and Frank, D.A. (1996). How can net primary productivity be measured in grazing ecosystems? *Ecology* 77: 974–7.

McNaughton, S.J., Oesterheld, M., Frank, D.A., and Williams, K.J. (1989). Ecosystem-level patterns of primary productivity and herbivory in terrestrial habitats. *Nature* 341: 142–4.

McNaughton, S.J. and J.L. Tarrants (1978). Grass leaf silicification: Natural selection for an inducible defense against herbivores. *Proceedings of the National Academy of Sciences* 80: 790–1.

Mduma, S.A.R., Sinclair, A.R.E., and Hilborn, R. (1999). Food regulates the Serengeti wildebeest: a 40-year record. *Journal of Animal Ecology* 68: 1101–22.

Milchunas, D.G. and Lauenroth, W.K. (1993). Quantitative effects of grazing on vegetation and soils over a global range of environments. *Ecological Monographs* 63: 327–66.

Milchunas, D.G., Sala, O.E., and W.K. Lauenroth (1988). A generalized model of the effects of grazing by large herbivores on grassland community structure. *The American Naturalist* 132: 87–106.

Moore, R.M. (1970). *Australian Grasslands*. Australia: Australian National University Press, p. 473.

Moran, M.D. (2014). Bison grazing increases arthropod abundance and diversity in a tallgrass prairie. *Environmental Entomology* 43: 1174–84.

Moran, M.D., Rooney, T.P., and L.E. Hurd (1996) Top-down cascade from a bitrophic predator in an old-field community. *Ecology* 77: 2219–27.

Moran, M.D. and Scheidler, A.R. (2002). Effects of nutrients and predators on an old-field food chain: interactions of top-down and bottom-up processes. *Oikos* 98: 116–24.

Morgan, J.A., LeCain, D.R., Pendall, E., Blumenthal, D.M., et al. (2011). C_4 grasses prosper as carbon dioxide eliminates desiccation in warmed semi-arid grassland. *Nature* 476: 202–5.

Mortensen, B., Danielson, B., Harpole, W.S., et al. (2018). Herbivores safeguard plant diversity by reducing variability in dominance. *Journal of Ecology* 106: 101–12. https://doi.org/10.1111/1365-2745.12821.

Morgan, J.A., Milchunas, D.G. LeCain, D.R., West, M., and A.R. Mosier (2007). Carbon dioxide enrichment alters plant community structure and accelerates shrub growth in the shortgrass steppe. *PNAS* 104: 14724–9.

Morgan, J.A., Mosier, A.R., Milchunas, D.G., LeCain, D.R., Nelson, J.A., and Parton, W.J. (2004). CO_2 enhances productivity, alters species composition, and reduces digestibility of shortgrass steppe vegetation. *Ecological Applications* 14: 208–19.

Mortensen, B., Danielson, B., Harpole, W. S., et al. (2018). Herbivores safeguard plant diversity by reducing variability in dominance. *Journal of Ecology* 106: 101–112. https://doi.org/10.1111/1365-2745.12821.

Mueller, K.E., Tilman, D., Fornara, D.A., and Hobbie, S.E. (2013). Root depth distribution and the diversity–productivity relationship in a long-term grassland experiment. *Ecology* 94: 787–93.

Mulder, C.P.H., Bazeley-White, E., Dimitrakopoulos, P. G., Hector, A., Scherer-Lorenzen, M., and Schmid, B. (2004). Species evenness and productivity in experimental plant communities. *Oikos* 107: 50–63.

Naeem, S., Thompson, L.J., Lawler, S.P., Lawton, J.H., and Woodfin, R.M. (1994). Declining biodiversity can alter the performance of ecosystems. *Nature* 368: 734–7.

Nelson, P.W. (1985). *The Terrestrial Natural Communities of Missouri*. Missouri Natural Areas Committee.

Nichols, E., Spectora, S., Louzadab, J., Larsen, T., Amezquitad, S., and M.E. Favilad (2008). 2008. Ecological functions and ecosystem services provided by *Scarabaeinae* dung beetles. *Biological Conservation* 141: 1461–74.

Nippert, J.B. and Knapp, A.K. (2007). Soil water partitioning contributes to species coexistence in tallgrass prairie. *Oikos* 116: 1017–29.

Nippert, J.B. and A.K. Knapp (2007). Linking water uptake with rooting patterns in grassland species. *Oecologia* 153: 261–72.

Nowacki, G.J. and M.D. Abrams (2008). The demise of fire and "mesophication" of forests in the Eastern United States. *Bioscience* 58: 123–38.

Oesterheld, M. and S.J. McNaughton (1991). Effect of stress and time for recovery on the amount of compensatory growth after grazing. *Oecologia* 85: 305–13.

Oesterheld, M. and Oyarzábal, M. (2004). Grass-to-grass protection from grazing in a semi-arid steppe. Facilitation, competition, and mass effect. *Oikos* 107(3): 576–82.

Oesterheld, M., Sala, O.E. and S.J. McNaughton (1992). Effect of animal husbandry on herbivore-carrying capacity at a regional scale. *Nature* 356: 234–6.

Pacala, S.W. and M. Rees (1998). Models suggesting field experiments to test two hypotheses explaining successional diversity. *The American Naturalist* 152: 729–37.

Paige, K.N. and T.G. Whitham (1987). Overcompensation in response to mammalian herbivory: The advantage of being eaten. *The American Naturalist* 129: 407–16.

Pan, J.J., Widner, B., Ammerman, D., and R.E. Drenovsky (2010). Plant community and tissue chemistry responses to fertilizer and litter nutrient manipulations in a temperate grassland. *Plant Ecology* 206: 139–50.

Parton, W.J., Morgan, J.A., Wang, G., and Del Grosso, S. (2007). Projected ecosystem impact of the Prairie heating and CO_2 enrichment experiment. *New Phytologist* 174: 823–34. doi:10.1111/j.1469-8137.2007.02052.x.

Pereira, J.A., Quintana, R.D., and S. Monge (2003). Diets of plains vizcacha, greater rhea and cattle in Argentina. *Journal of Range Management* 56: 13–20.

Peters, D.P.C., R.A. Pielke, B.T. Bestelmeyer, C.D. Allen. S. Munson-McGee, and K.M. Havstad (2004). Cross-scale interactions, nonlinearities, and forecasting catastrophic events. *PNAS* 101:15130–5.

Pianka, E.R. (1967). On lizard species diversity: North American flatland deserts. *Ecology* 48: 333–51. doi:10.2307/1932670.

Pimm, S. (1991). *The Balance of Nature?* Chicago: University of Chicago Press.

Pimm, S.L. (1984). The complexity and stability of ecosystems. *Nature* 307(5949): 321–6.

Platt, W.J. (1975). The colonization and formation of equilibrium plant species associations on badger disturbances in a tall-grass prairie. *Ecological Monographs* 45: 285–305.

Pleasants, J.M. Zalucki, M.P., Oberhauser, K.S., Brower, L.P., Taylor, O.R., and W.E. Throgmartin (2017). Interpreting surveys to estimate the size of the monarch butterfly population: Pitfalls and prospects. *PlosONE* doi.org/10.1371/journal.pone.0181245

Polley, H.W. and S.L. Collins (1984). Relationships of vegetation and environment in buffalo wallows. *American Midland Naturalist* 112: 178–86.

Polley, H.W., Derner, J.D., Jackson, R.B., Wilsey, B.J. and P.A. Fay (2014). Impacts of climate change drivers on C_4 grassland productivity: scaling driver effects through the plant community. *Journal of Experimental Botany* 13: 3415–24.

Polley, H.W., Johnson, D.M., and R.B. Jackson (2016). Canopy foliation and area as predictors of mortality risk from episodic drought for individual trees of Ashe juniper. *Plant Ecology* 217: 1105–14.

Polley, H.W., Mayeux, H.S., Johnson, H.B., and Tischler, C.R. (1997). Viewpoint: atmospheric CO_2, soil water, and shrub/grass ratios on rangelands. *Journal of Range Management* 278–84.

Polley, H.W., B.J. Wilsey, and J.D. Derner (2003). Do species evenness and plant density influence the magnitude of selection and complementary effects in annual plant mixtures? *Ecology Letters* 6: 248–57.

Polley, H.W., Wilsey, B.J., and J.D. Derner (2005). Patterns of plant species diversity in remnant and restored tallgrass prairies. *Restoration Ecology* 13: 480–7.

Polley, H.W., Wilsey, B.J., and Derner, J.D. (2007). Dominant species constrain effects of species diversity on temporal variability in biomass production of tallgrass prairie. *Oikos* 116: 20144–52.

Polley, H.W., Wilsey, B.J., Derner, J.D., Johnson, H.B. and J. Sanabria (2006). Early-successional plants regulate grassland productivity and species composition: a removal experiment. *Oikos* 113: 287–95.

Ponce-Campos, G.E., Moran, M.S., Huerte, A., Zhang, Y., and C. Bresloff (2013). Ecosystem resilience despite large-scale altered hydroclimatic conditions. *Nature* 494: 349–52.

Prach, K. and R.J. Hobbs (2008). Spontaneous succession versus technical reclamation in the restoration of disturbed Sites. *Restoration Ecology* 16: 363–6.

Preisser, E.L., Bolnick, D.I., and Benard, M.F. (2005). Scared to death? The effects of intimidation and consumption in predator-prey interactions. *Ecology* 86: 501–9. doi:10.1890/04-0719.

Prevéy, J.S. and Seastedt, T.R. (2014). Seasonality of precipitation interacts with exotic species to alter composition and phenology of a semi-arid grassland. *Journal of Ecology* 102: 1549–61. doi:10.1111/1365-2745.12320.

Price, J.N. and J.W. Morgan (2008). Woody plant encroachment reduces species richness of herb-rich woodlands in southern Australia. *Austral Ecology* 33: 278–89. doi:10.1111/j.1442-9993.2007.01815.x.

Pringle, A., J.D. Bever, M. Gardes, J.L. Parrent, M.C. Rillig, and J.N. Klironomos (2009). Mycorrhizal symbioses and plant invasions. *Annu. Rev. Ecol. Evol. Syst.* 40: 699–715.

Pyšek, P., Manceur, A.M., Alba, C., McGregor, K.F., Pergl, J., Stajerová, K., et al. (2014). Naturalization of central European plants in North America: Species traits, habitats, propagule pressure, residence time. *Ecology* 96: 762–74.

Questad, E.J. and Foster, B.L. (2008). Coexistence through spatio-temporal heterogeneity and species sorting in grassland plant communities. *Ecology Letters* 11: 717–26.

Reader, R., Wilson, S., Belcher, J., Wisheu, I., Keddy, P., Tilman, D., Morris, E.C., Grace, J.B., McGraw, J.B., Olff, H., Turkington, R., Klein, E., Leung, Y., Shipley, B., van Hulst, R., Johansson, M.E., Nilsson, C., Gurevitch, J., Grigulis, K., and Beisner, B. (1994). Plant Competition in relation to neighbor biomass: An intercontinental study with *Poa pratensis*. *Ecology* 75(6): 1753–60.

Raich, J.W. and W.H. Schlesinger (1992). The global carbon dioxide flux in soil respiration and its relationship to vegetation and climate. *Tellus B* 44: 81–99. doi:10.1034/j.1600-0889.1992.t01-1-00001.x.

Ratnam J., et al. (2011). When is a 'forest' a savanna, and why does it matter? *Global Ecology and Biogeography* 20: 653–60.

Reed, et al. (2018). Testing the cultivar vigor hypothesis: comparisons of the competitive ability of wild and cultivated populations of *Pascopyrum smithii* along a restoration chronosequence. *Restoration Ecology* in press.

Reich, P.B., Knops, J., Tilman, D., Craine, J., Ellsworth, D., et al. (2001). Plant diversity enhances ecosystem responses to elevated CO_2 and nitrogen deposition. *Nature* 410: 809–10.

Richardson, D.M., Pyšek, P., Rejmánek, M., Barbour, M.G., Panetta, F.D., and West, C.J. (2000). Naturalization and invasion of alien plants: concepts and definitions. *Diversity and Distributions* 6: 93–107. doi:10.1046/j.1472-4642.2000.00083.x.

Ricketts, T.H., Dinerstein, E., Olson, D.M., and C.J. Loucks, et al. (1999). *Terrestrial Ecoregions of North America*. Washington: Island Press.

Ripple, W.J. and R.L. Beschta (2004). Wolves, elk, willows, and trophic cascades in the upper Gallatin Range of Southwestern Montana, USA. *Forest Ecology and Management* 200: 161–81.

Risch, A. and D.A. Frank (2005). Carbon dioxide fluxes in a spatially and temporally heterogeneous temperate grassland. *Oecologia*, 147: 291–302.

Rivrud, I.M., Heurich, M., Krupczynski, P., Müller, J., and Mysterud, A. (2016). Green wave tracking by large herbivores: An experimental approach. *Ecology* 97: 3547–53.

Root, R.B. (1973). Organization of a plant–arthropod association in simple and diverse habitats: the fauna of collards (*Brassica oleracea*). *Ecological Monographs* 43: 95–124.

Rosenzweig, M.L. (1995). *Species diversity in space and time*. Cambridge: Cambridge University Press.

Ruimy, A., Jarvis, P.G., Baldocchi, D.D., and Saugier, B. (1995). CO_2 fluxes over plant canopies and solar radiation: a review. *Advances in Ecological Research* 26: 1–68.

Running, S.W. (2012). A measurable planetary boundary for the biosphere. *Science* 337: 1458–9.

Sage, R.F. (2004). The evolution of C_4 photosynthesis. *New Phytologist* 161: 341–70. doi:10.1111/j.1469-8137.2004.00974.x.

Sala, O.E., Gherardi, L.A., Reichmann, L., Jobbagy, E., and D. Peters (2012). Legacies of precipitation fluctuations on primary production: theory and data synthesis. *Phil. Trans. R. Society London B. Biol. Sci.* 367: 31335–44.

Samson, F. and Knopf, F. (1994). Prairie Conservation in North America. *BioScience* 44(6). 418–21.

Sankaran, M., Hanan, N.P., et al. (2005). Determinants of woody cover in African savannas. *Nature* 438: 846–9.

Sasaki, T. and Lauenroth, W.K. (2011). Dominant species, rather than diversity, regulates temporal stability of plant communities. *Oecologia* 166: 761–8.

Savage, V.M., Gillooly, J.F., Woodruff, W.H., West, G.B., Allen, A.P., Enquist, B.J. and Brown, J.H. (2004). The predominance of quarter-power scaling in biology. *Functional Ecology* 18: 257–82.

Schlesinger, W.H. (1994). *Biogeochemistry: An Analysis of Global Change.* New York and London: Academic Press.

Schlesinger, W.H., Reynolds, J.F., Cunningham, G.L., Huenneke, L.F., Wesley, M., Jarrell, W. M., Ross A. Virginia, R.A., and W.G. Whitford (1990). Biological feedbacks in global desertification. *Science* 247: 1043–8.

Scholes, R.J. and D.O. Hall (1996). The carbon budget of tropical savannas, woodlands and grasslands. In *Global Change: Effects on Coniferous Forests and Grasslands*, SCOPE. Chichester: Wiley.

Schmitz, O.J. (2003). Top predator control of plant biodiversity and productivity in an old-field ecosystem. *Ecology Letters* 6: 156–63. doi:10.1046/j.1461-0248.2003.00412.x.

Schnitzer, S.A., Klironomos, J.N., HilleRisLambers, J., Kinkel, L.L., Reich, P.B., Xiao, K., Rillig, M.C., Sikes, B.A., Callaway, R.M., Mangan, S.A., Van Nes, E.H., and Scheffer, M. (2011). Soil microbes drive the classic plant diversity-productivity pattern. *Ecology* 92: 296–303.

Schulte, L.A., Niemi, J., Helmers, M.J., Liebman, M., Arbuckle, J.G., et al. (2017). Prairie strips improve biodiversity and the delivery of multiple ecosystem services from corn–soybean croplands. *PNAS* 114:11249–52. doi:10.1073/pnas.1620229114.

Schulze, E.D. and H.A. Mooney (1994). *Biodiversity and Ecosystem Function.* Berlin, Heidelberg: Springer.

Schlesinger, W.H., Reynolds, J.F., Cunningham, G.L., Huenneke, L.F., Wesley M., Jarrell, W.M., Ross A. Virginia, R.A., and W.G. Whitford (1990). Biological feedbacks in global desertification. *Science* 247: 1043–8.

Schwilk, D.W. and N. Zavala. (2012). Germination response of grassland species to plant-derived smoke. *Journal of Arid Environments* 79: 111–15.

Scurlock, J.M.O., Johnson, K., and Olson, R.J. (2002). Estimating net primary productivity from grassland biomass dynamics measurements. *Global Change Biology* 8: 736–53.

Sensenig, R.L., Demment, M.W., and Laca, E.A. (2010). Allometric scaling predicts preferences for burned patches in a guild of East African grazers. *Ecology* 91: 2898–907. doi:10.1890/09-1673.1.

Silvertown, J., Poulton, P., Johnston, E., Edwards, G., Heard, M., and Biss, P.M. (2006). The Park Grass Experiment 1856–2006: its contribution to ecology. *Journal of Ecology* 94(4): 801–14.

Sinclair, A.R.E. and Arcese, P. (eds.) (1995). *Serengeti II: Dynamics, Management, and Conservation of an Ecosystem.* Chicago: University of Chicago Press.

Sinclair, A.R.E., Mduma, S.A.R., and Arcese, P. (2000). What determines phenology and synchrony of ungulate breeding in Serengeti? *Ecology* 81: 2100–111.

Sluis, W.J. (2002). Patterns of species richness and composition in re-created grassland. *Restoration Ecology* 10: 677–84.

Smith, M.D. and Knapp, A.K. (2003). Dominant species maintain ecosystem function with non-random species loss. *Ecology Letters* 6(6): 509–17.

Smith, M.D., Knapp, A.K., and Collins, S.L. (2009). A framework for assessing ecosystem dynamics in response to chronic resource alterations induced by global change. *Ecology* 90: 3279–89.

Sneath, D. (1998). State policy and pasture degradation in Inner Asia. *Science* 281: 1147–8. 10.1126/science.281.5380.1147.

Soriano, A. (1991). Rio de la Plata grasslands. Ecosystems of the world. Natural grasslands. Chapter 19 in Coupland R.T. (ed.) *Agriculture, Ecosystems & Environment.* Amsterdam: Elsevier, pp. 367–407.

Srivastava, D.S. and Vellend, M. (2005). Biodiversity-ecosystem function research: is it relevant to conservation? *Annual Review of Ecology, Evolution, and Systematics* 36: 267–94.

Stampfli, A. and M. Zeiter (2004). Plant regeneration directs changes in grassland composition after extreme drought: a 13-year study in southern Switzerland. *Journal of Ecology* 92: 568–76.

Steinauer, E.M. and Collins, S.L. (1995). Effects of urine deposition on small-scale patch structure in prairie vegetation. *Ecology* 76: 1195–205. doi:10.2307/1940926.

Still, C.J., J.A. Berry, G.J. Collatz, and R.S. DeFries (2003). Global distribution of C_3 and C_4 vegetation: Carbon cycle implications, *Global Biogeochemistry Cycles* 17: 1006: doi:10.1029/2001GB001807.

Stirling, G. and Wilsey, B. (2001). Empirical relationships between species richness, evenness, and proportional diversity. *The American Naturalist* 158: 286–99.

Stiling, P., Rossi, A.M., Hungate, B., Dijkstra, P., Hinkle, C.R., Knott, W.M., and Drake, B. (1999). Decreased leaf-miner abundance in elevated CO_2: reduced leaf quality and increased parasitoid attack. *Ecological Applications* 9: 240–4. doi:10.1890/1051-0761(1999)009[0240:DLMAIE]2.0.CO;2.

Strauss, S.Y. and A.A. Agrawal (1999). The ecology and evolution of plant tolerance to herbivory. *Trends in Ecology and Evolution* 14: 179–85.

Stronach, N.R.H. and S.J. McNaughton (1989). Grassland fire dynamics in the Serengeti Ecosystem, and a potential method of retrospectively estimating fire energy. *Journal of Applied Ecology* 26: 1025–33.

Stuble, K.L., S.E. Frick, and T.P. Young (2017). Every restoration is unique: testing year effects and site effects as drivers of initial restoration trajectories. *Journal of Applied Ecology* 54(4). doi:10.1111/1365-2664.12861.

Suding, K.N., Collins, S.L., Gough, L., Clark, C., Cleland, E.E., Gross, K.L., Milchunas, D.G., and Pennings, S. (2005). Functional-and abundance-based mechanisms explain diversity loss due to N fertilization. *Proceedings of the National Academy of Sciences of the United States of America* 102(12): 4387–92.

Suding K.N., Gross, K.L., and Houseman, G.R. (2004). Alternative states and positive feedbacks in restoration ecology. *Trends in Ecology and Evolution* 19: 46–53.

Sun, G., Coffin, D.P., and Lauenroth, W.K. (1997). Comparison of root distributions of species in North American grasslands using GIS. *Journal of Vegetation Science* 8: 587–96.

Swan, C.M., Gluth, M.A., and Horne, C.L. (2009). Leaf litter species evenness influences nonadditive breakdown in a headwater stream. *Ecology* 90: 1650–8. doi:10.1890/08-0329.1.

Tansley A.G. and Adamson R.S. (1925). Studies of the vegetation of the English chalk: III the chalk grasslands of the Hampshire–Sussex border. *Journal of Ecology* 13: 177–223.

Taylor L.R. (1961). Aggregation, variance and the mean. *Nature*, 189, 732–5.

Teeri, J.A. and Stowe, L.G. (1976). Climatic patterns and the distribution of C_4 grasses in North America. *Oecologia* 23: 1. https://doi.org/10.1007/BF00351210.

Temperton, V.M. and R.J. Hobbs (2004). The search for ecological assembly rules and its relevance to restoration ecology. In Temperton, V.M., et al. (ed.) *Assembly Rules and Restoration Ecology*. New York: Island Press.

Tilman, D. (1982). Resource Competition and Community Structure. *Monogr. Pop. Biol. 17.* Princeton, NJ: Princeton University Press, p. 296.

Tilman, D. (1999). The ecological consequences of changes in biodiversity: a search for general principles. *Ecology* 80: 1455–74.

Tilman, D. and Downing, J.A. (1994). Biodiversity and stability in grasslands. *Nature* 367: 363–5.

Tilman, D., Knops, J., Wedin, D., Reich, P., Ritchie, M., and E. Siemann (1997). The influence of functional diversity and composition on ecosystem processes. *Science* 277: 1300–2.

Tilman, D., Reich, P.B., and Knops, J.M. (2006). Biodiversity and ecosystem stability in a decade-long grassland experiment. *Nature* 441: 629–32.

Tilman, D., Reich, P.B., Knops, J., Wedin, D., Mielke, T., and C. Lehman (2001). Diversity and productivity in a long-term grassland experiment. *Science* 294: 843–5.

Tilman, D. and Wedin, D. (1991). Dynamics of nitrogen competition between successional grasses. *Ecology* 72(3): 1038–49.

Tognetti, P.M., Chaneton, E.J., M. Omacini, H.J. Trebino, and R.J.C. León (2010). Exotic vs. native plant dominance over 20 years of old-field succession on set-aside farmland in Argentina. *Biol. Conserv.* 143: 2494–503.

Towne, E., Hartnett, D.C., and Cochran, R.C. (2005). Vegetation trends in tallgrass prairie from bison and cattle grazing. *Ecological Applications* 15: 1550–9.

Towne, E.G. (2000). Prairie vegetation and soil nutrient responses to ungulate carcasses. *Oecologia* 122: 232–9.

Van Auken, O.W. (2000). Shrub invasions of North American semiarid grasslands. *Annual Review of Ecology and Systematics* 31: 197–215.

van der Heijden, M.G.A., Bardgett, R.D., and Van Straalen, N.M. (2008). The unseen majority: soil microbes as drivers of plant diversity and productivity in terrestrial ecosystems. *Ecology Letters* 11: 296–310.

Van Soest, P.J. (1982). *Nutritional Ecology of The Ruminant.* Cornell: Cornell University Press.

Veldman, J.W., Overbeck, G.E., Negreiros, D., Mahy, G., et al. (2015). Where tree planting and forest expansion are bad for biodiversity and ecosystem services, *BioScience* 65: 1011–18.

Vitousek, P.M. (1994). Beyond global warming: ecology and global change. *Ecology* 75: 1861–76. doi:10.2307/1941591.

Vitousek, P.M., Aber, J.D., Howarth, R.W., Likens, G.E., Matson, P.A., Schindler, D.W., Schlesinger, W.H., and Tilman, D.G. (1997). Human alteration of the global nitrogen cycle: sources and consequences. *Ecological applications* 7: 737–50.

Vitousek, P.M., Ehrlich, P.R., Ehrlich, A.H., and P.A. Matson (1986). Human Appropriation of the Products of Photosynthesis. *Bioscience* 36: 368–73.

Vitousek, P.M. and Hooper, D.U. (1993). Biological diversity and terrestrial ecosystem biogeochemistry. In Schulze, E.-D. and Mooney, H.A. (eds.) *Biodiversity and Ecosystem Function*. Berlin: Springer, p. 3.

Wainwright, C.E., Wolkovich, E.M., and Cleland, E.E. (2012). Seasonal priority effects: implications for invasion and restoration in a semi-arid system. *Journal of Applied Ecology* 49: 234–41. doi:10.1111/j.1365-2664.2011.02088.x.

Waller, D. and Alverson, W. (1997). The White-Tailed Deer: A Keystone Herbivore. *Wildlife Society Bulletin (1973–2006)* 25(2): 217–26.

Wand, S.J.E., Midgley, Guy. F., Jones, M.H., and Curtis, P.S. (1999). Responses of wild C_4 and C_3 grass (*Poaceae*) species to elevated atmospheric CO_2 concentration: a meta-analytic test of current theories and perceptions. *Global Change Biology* 5: 723–41. doi:10.1046/j.1365-2486.1999.00265.x.

Wang, D. and Ba, L. (2008). Ecology of meadow steppe in northeast China. *The Rangeland Journal* 30: 247–54.

Wang, L., Wang, D., He, Z., Liu, G., and Hodgkinson, K.C. (2010). Mechanisms linking plant species richness to foraging of a large herbivore. *Journal of Applied Ecology* 47: 868–75. doi:10.1111/j.1365-.

Wardle, D.A., Bardgett, R.D., Callaway, R.M., and Van der Putten, W.H. (2011). Terrestrial ecosystem responses to species gains and losses. *Science* 332: 1273–7.

Wardle, D.A., Bonner, K.I., and Nicholson, K.S. (1997). Biodiversity and plant litter: experimental evidence which does not support the view that enhanced species richness improves ecosystem function. *Oikos* 79(2): 247–58.

Weaver, J.E. (1954). *North American Prairie*. Lincoln, Nebraska: Johnsen Publishing.

Weaver, J.E. (1968). *Prairie Plants and Their Environment*. Lincoln, Nebraska: University of Nebraska.

Weber, E. (2017). *Invasive Plant Species of the World*. Second ed. EU: CABI, p. 568.

Wedin, D.A. and D. Tilman (1996). Influence of nitrogen loading and species composition on the carbon balance of grasslands. *Science* 274: 1720–3.

Weiher, E., Forbes, S., Schauwecker, T., and B. Grace, J. (2004). Multivariate control of plant species richness and community biomass in blackland prairie. *Oikos* 06: 151–7.

Weiher, E. and P. Keddy (1999). *Ecological Assembly Rules*. Cambridge: Cambridge University Press.

Weltzin, J.F., Lolk, M.E., Schwinning, S., Williams, D.G., and P.A. Fay (2003). Assessing the response of terrestrial ecosystems to potential changes in precipitation. *Bioscience* 53: 941–52.

Weltzin, J.F., Archer, S., and Heitschmidt, R.K. (1997). Small-mammal regulation of vegetation structure in a temperate savanna. *Ecology* 78: 751–63. doi:10.1890/0012-9658(1997)078[0751:SMROVS]2.0.CO;2.

Westoby, M. (1998). A leaf-height-seed (LHS) plant ecology strategy scheme. *Plant and Soil* 199: 213–27.

Westoby, M., B. Walker., and I. Noy-Meir (1989). Opportunistic management for rangelands not at equilibrium. *Journal of Range Management* 42: 266–74.

White, F. (1983). *The vegetation of Africa: A descriptive memoir to accompany the UNESCO/AETFAT/UNSO, Vegetation map of Africa*. Switzerland: United Nations, UNESCO.

White, P.S. and Walker, J.L. (1997). Approximating nature's variation: selecting and using reference information in restoration ecology. *Restoration Ecology* 5: 338–49.

Whittaker, R.H. (1960). Vegetation of the Siskiyou Mountains, Oregon and California. *Ecological Monographs* 30: 279–338.

Williams, C.B. (1964). *Patterns in the Balance of Nature and Related Problems of Quantitative Ecology*. London and New York: Academic Press.

Williams, K.J., Wilsey, B.J., McNaughton, S.J., and F. Banyikwa (1998). Temporal variation in rainfall does not decrease the yield of Serengeti grasses. *Oikos* 81: 463–73.

Wilmers, C.C., Stahler, D.R., Crabtree, R.L., Smith, D.W., and Getz, W.M. (2003). Resource dispersion and consumer dominance: scavenging at wolf- and hunter-killed carcasses in Greater Yellowstone, USA. *Ecology Letters* 6: 996–1003. doi:10.1046/j.1461-0248.2003.00522.x.

Wilsey, B.J. (1996). Variation in use of green flushes following burns among African ungulate species: the importance of body size. *African Journal of Ecology* 34: 32–8.

Wilsey, B.J. (2010). Productivity and subordinate species response to dominant grass species and seed source during restoration. *Restoration Ecology* 18: 628–37.

Wilsey, B.J., Barber, K., and L.M. Martin (2015). Exotic grassland species have stronger priority effects than natives regardless of whether they are cultivated or wild genotypes. *New Phytologist* 205: 928–37.

Wilsey, B.J., Chalcraft, D.R., Bowles, C.M., and Willig, M.R. (2005). Relationships among indices suggest that richness is an incomplete surrogate for grassland biodiversity. *Ecology* 86: 1178–84.

Wilsey, B.J., Coleman, J.S., and McNaughton, S.J. (1997). Effects of elevated CO_2 and defoliation on grasses: a comparative ecosystem approach. *Ecological Applications* 7(3): 844–53.

Wilsey, B.J., P.P. Daneshgar, and H.W. Polley (2011). Biodiversity, phenology and temporal niche differences between native- and novel exotic-dominated grasslands. *Perspectives in Plant Ecology, Evolution and Systematics* 13: 265–76.

Wilsey, B.J., Daneshgar, P.P., Hofmockel, K., and Polley, H.W. (2014). Invaded grassland communities have altered stability-maintenance mechanisms but equal stability compared to native communities. *Ecology Letters* 17: 92–100.

Wilsey, B.J. and L.M. Martin (2015). Top-down control of rare species abundances by native ungulates in a grassland restoration. *Restoration Ecology* 23: 465–72.

Wilsey, B.J, L.M. Martin, and A.D. Kaul (2017). Phenology differences between native and novel exotic-dominated grasslands rival the effects of climate change. *Journal of Applied Ecology* in press. doi:10.1111/1365-2664-1291.

Wilsey, B.J., Martin, L.M. and H.W. Polley (2005). Predicting plant extinction based on species-area curves in prairie fragments with high beta richness. *Conservation Biology* 19: 1835–41.

Wilsey, B.J., McNaughton, S.J., and Coleman, J.S. (1994). Will increases in atmospheric CO_2 affect regrowth following grazing in C_4 grasses from tropical grasslands? A test with *Sporobolus kentrophyllus*. *Oecologia* 99: 141–4.

Wilsey, B.J., G. Parent, N.T. Roulet, T.R. Moore, and C. Potvin (2002). Tropical pasture carbon cycling: relationships between C source/sink strength, aboveground biomass, and grazing. *Ecology Letters* 5: 367–76.

Wilsey, B.J. and H.W. Polley (2002). Reductions in grassland species evenness increase dicot seedling invasion and spittle bug infestation. *Ecology Letters* 5: 676–84.

Wilsey, B.J. and H.W. Polley (2003). Effects of seed additions and grazing history on diversity and aboveground productivity of sub-humid grasslands. *Ecology* 84: 920–31.

Wilsey, B.J., and H.W. Polley (2004). Realistically low species evenness does not alter grassland species-richness-productivity relationships. *Ecology* 85: 2693–700.

Wilsey, B.J. and H.W. Polley (2006). Aboveground productivity and root-shoot allocation differ between native and introduced grass species. *Oecologia* 150: 300–9.

Wilsey, B.J. and Potvin, C. (2000). Biodiversity and ecosystem functioning: importance of species evenness in an old field. *Ecology* 81: 887–92.

Wilsey, B. and G. Stirling (2007). Species richness and evenness respond in a different manner to propagule density in developing prairie microcosm communities. *Plant Ecology* 190: 259–73.

Wilsey, B.J., Teaschner, T.B., Daneshgar, P.P., Isbell, F.I., and H. W. Polley (2009). Biodiversity maintenance mechanisms differ between native and novel exotic-dominated communities. *Ecology Letters* 12: 432–42.

Wilson, J.B. (1999). Assembly rules in plant communities. In E. Weiher and P. Keddy (eds.) *Ecological Assembly Rules: Perspectives, Advances, Retreats.* Cambridge: Cambridge University Press.

Wilson, E.O. (1992). *The Diversity of Life.* New York: W.W. Norton & Company.

Wolkovich, E.M. and Cleland, E.E. (2011). The phenology of plant invasions: A community ecology perspective. *Frontiers in Ecology and the Environment* 9: 287–94.

Wright, H.A. (1974). Effect of fire on southern mixed prairie grasses. *Journal of Range Management* 27: 417–19.

Xia, J.Y., et al. (2015). Joint control of terrestrial gross primary productivity by plant phenology and physiology. *Proceedings of the National Academy of Sciences* 112: 2788–93.

Xia, L., Yang, Q., Li, Z., Wu, Y., and Feng, Z. (2007). The effect of the Qinghai-Tibet railway on the migration of Tibetan antelope *Pantholops hodgsonii* in Hoh-xil National Nature Reserve, China. *Oryx* 41: 352–7. doi:10.1017/S0030605307000116.

Xu, X., H.W. Polley, K. Hofmockel, and B.J. Wilsey (2017). Species composition but not diversity explains recovery from the 2011 drought in Texas grasslands. *Ecosphere* 8: e01704.

Young, T.P., Petersen, D.A., and Clary, J.J. (2005). The ecology of restoration: historical links, emerging issues and unexplored realms. *Ecology Letters* 8: 662–73.

Young, T.P., K.L. Stuble, K.L., J.A. Balachowski, and C.M. Werner (2017). Using priority effects to manipulate competitive relationships in restoration. *Restoration Ecology*, online early.

Yurkonis, K.A., Meiners, S.J., and Wachholder, B.E. (2005). Invasion impacts diversity through altered community dynamics. *Journal of Ecology* 93: 1053–61. doi:10.1111/j.1365-2745.2005.01029.x.

Yurkonis, K.A. Wilsey, B.J., and K.A. Moloney (2012). Initial plant arrangement affects invasion resistance in experimental grassland plots. *Journal of Vegetation Science* 23: 4–12.

Zavaleta, E.S. and Hulvey, K.B. (2004). Realistic species losses disproportionately reduce grassland resistance to biological invaders. *Science* 306(5699): 1175–7.

Zavaleta, E.S., Pasari, J.R., Hulvey, K.B., and Tilman, G.D. (2010). Sustaining multiple ecosystem functions in grassland communities requires higher biodiversity. *Proceedings of the National Academy of Sciences* 107(4): 1443–6.

Zhong Z, Li X, Pearson D, Wang D, Sanders D, Zhu Y, and Wang L. (2017). Ecosystem engineering strengthens bottom-up and weakens top-down effects via trait mediated indirect interactions. *Proceedings of the Royal Society B* 284: 20170475. http://dx.doi.org/10.1098/rspb.2017.0894.

Zuppinger-Dingley, D., B. Schmid, J.S. Petermann, V. Yadav, G.B. De Deyn, and D.F.B. Flynn (2014). Selection for niche differentiation in plant communities increases biodiversity effects. *Nature* 515: 108–11.

Index